PATTERNS
in the
SAND

FRONTIERS OF SCIENCE
Series editor: Paul Davies

Forthcoming:

COMPUTERS, COMPLEXITY, AND EVERYDAY LIFE

PATTERNS
in the
SAND

TERRY BOSSOMAIER • DAVID GREEN

FOREWORD BY PAUL DAVIES

§

HELIX BOOKS

PERSEUS BOOKS
Reading, Massachusetts

Many of the designations used by manufacturers and sellers to distinguish their products are claimed as trademarks. Where those designations appear in this book and Perseus Books was aware of a trademark claim, the designations have been printed in initial capital letters.

Library of Congress Catalog Card Number: 98–86413

ISBN 0-7382-0015-8

Published in Australia by Allen & Unwin Pty Ltd

Perseus Books is a member of the Perseus Books Group

Jacket design by Suzanne Heiser
Set in 10.5-point Plantin by DOCUPRO, Sydney

1 2 3 4 5 6 7 8 9–DOH–0201009998
First printing, July 1998

Find Helix Books on the World Wide Web at
http://www.aw.com/gb/

CONTENTS

FOREWORD

A casual glance at the world around us reveals such a rich diversity of physical forms and systems it seems incredible that we could ever come to understand nature. The Greek philosophers struggled with this challenge and one of their number, Democritus, proposed a neat solution. The entire universe, he maintained, was composed of nothing but atoms moving in a void. In this way, the huge variety of physical things might be attributed to the unlimited number of ways in which these atoms could combine and, therefore, everything that happens in nature would be due to the rearrangement of these atoms. The immense complexity of the physical universe could then be reduced to the antics of a few species of basic building blocks.

The reductionist world view that stemmed from Democritus' atomism was destined to become a paradigm for science. Nature may be bewilderingly complex, it was said, but if only the nature of the building blocks could be understood, then all would be explained, for the whole would be nothing more than the sum of its parts. It is an enticing philosophy, however, reductionism has been roundly condemned for committing the fallacy of 'nothing-buttery'. Take a human being. Since people are made up of cells, the reductionist seeks to explain human behaviour by understanding how cells (including brain cells) work. But cells are just complicated bags of molecules, so their properties should be explicable by chemistry. Chemistry, in turn, is just a branch of physics, so we should, in principle, be able to derive the chemical properties of bio-molecules from the laws of atomic and subatomic physics. Thus psychology and biology reduce to chemistry, which in turn reduces to particle physics. If this chain of reasoning were correct, it would mean that, ultimately, all human qualities could be embedded in the properties of the underlying building blocks—the fundamental particles and fields out of which the world is put together.

While reductionism remains a powerful methodology, few scientists believe it is the whole story. Over the last century, physics has cured itself of its once strong reductionist flavour, as more and more phenomena have been discovered that don't fit comfortably into the paradigm. Historically, one can see glimmerings of trouble for reductionism in the work of Henry Poincaré, a French mathematical physicist who, at the turn of the century, discovered what we would now call the theory of chaos. Before this, it was generally supposed that complexity is mere complication caused by the sheer number of participating units. No fundamentally new laws or principles were expected to be involved. Thus, a biological organism or a snowflake or a turbulent stream may appear complex, but this is only because we do not know the details of all the tiny components that constitute them nor the way that they interact. It was generally supposed that beneath the surface complication of such systems, there must lie simplicity at the level of individual components. Chaos theory demolished this belief by showing that even simple systems with very few components can be so exquisitely sensitised to the tiniest of disturbances that they behave in an essentially random and unpredictable manner.

But this was only the beginning. Other forms of complexity, where the random anarchy of chaos is replaced by more constructive activity, were discovered. The laser and the superconductor showed that countless independent elements can act collectively in a cooperative manner, as if guided by an unseen hand. Even a humble pan of hot water can spontaneously arrange itself into a pattern of hexagonal convection cells. The terms 'synergetics' and 'self-organisation' were coined to describe the coherent activity of complex systems that display long-range cooperative behaviour. Sometimes assemblages of simple, unexciting components will take on a life of their own and act in surprising and even creative ways that are impossible to predict. Often small external changes will trigger major consequences as a system reaches an instability that may lead either to a catastrophe or mark the transition to a state of still more elaborate complexity.

But how can scientists understand these complex systems? The problem about complexity is that, by its very nature, it seems impossible to grasp in generality. No two cloud patterns, planetary rings or biological organisms are the same. However, with the advent of the electronic computer, researchers were able to study a vast range of complex behaviour. They started to notice similarities in the structure and activity of widely different complex systems. Universal mathematical patterns were discerned even in chaotic systems. It began to look as if there might be quite general laws of complexity to place alongside the laws of physics.

A completely new culture began to emerge among scientists studying such complex systems. The reductionist paradigm is founded on the idea of treating the smallest components as primitive, irreducible and simple, then trying to explain complex systems by the combined activity of their myriad parts. Complexity theorists turned this on its head. Instead of a bottom-up approach, they sought examples of primitive, irreducibly complex components. One such is the fractal, a geometrical object that is in some sense infinitely irregular (just as Democritus' atoms were infinitely simple). Nevertheless, fractals have a very elegant and manageable mathematical description and so can be used to model complexity very efficiently. An often-cited example is the coastline of Great Britain, which is wiggly on all scales of size, but in a way that has a simple underlying scaling relation. Indeed, it is often cheaper on data to fractalise the image of a natural scene rather than to pixelate it (which is the reductionist's approach). This is because nature is full of fractal patterns—not only coastlines, but also mountain ranges, forests and river valleys, to name but a few.

Another important concept is the network—a collection of objects connected in a way that they can influence each other to a greater or lesser extent. Networks are familiar from the structure of brains, the evolution of ecosystems and the movements of markets. They occur in magnetic materials and immune systems, in ant colonies and gene interactions and, most famously, in the Internet and World Wide Web. By definition, a network cannot be reduced to its components, for it is its very inter-connectedness that bestows upon a network its distinctive qualities. Once again, researchers see a hint of universal principles at work. They glimpse self-organisation and chaos in systems as diverse as currency markets and rainforests and the patterns that decorate the surfaces of sand piles. Their work holds out the tantalising promise that aspects of nature (or even human behaviour) which once appeared incomprehensible and utterly intractable might yet be accurately modelled using new mathematical principles.

Terry Bossomaier and David Green belong to this new breed of researcher. Their expertise transcends traditional subject boundaries. Together, they bring to the study of complexity a background of biology, computing, neural nets and mathematics. While sceptical of some of the wilder claims made for complexity theory, they are convinced that we stand on the threshold of a new science that promises to have wide-ranging implications for our understanding of nature. In this book, they describe many everyday examples of the complexity theorist's craft at work. Their account leaves no doubt that the subject is not just one of grandiose visions and paradigm-busting leaps: it is also about down-to-earth practical applications. Take, for example, their discussion of accidents and criminal responsibility, or the spread of epidemics—both cases where increasing social complexity can cause escalating problems and instabilities.

When I was a student, it was often said that the two great frontiers of science were the very large and the very small—cosmology and particle physics. Today there is a third frontier, equally challenging and important, that of the very complex. There are many occasions in the history of science where the word 'revolution' has been glibly applied. True scientific revolutions are actually rather rare, however, and involve more than sudden advances in this or that discipline. The complexity revolution is aptly named, for it goes beyond merely providing interesting examples: it changes the way that scientists approach their subject matter. It has given us new tools and concepts, as well as fresh explanations for age-old puzzles. Mankind has always known that the physical universe is complex—magnificently, exuberantly, richly complex—but only now are we beginning to understand just how subtle and extensive that complexity may be.

Paul Davies
Adelaide, 1997

FIGURES

xi

GLOSSARY

algorithm A recipe for performing a function or solving a problem.

alife (artificial life) The study of life-like properties in artificial systems.

analog/digital Analog processes use 'analogies' of physical phenomena (e.g., electrical voltage for the size of a variable), whereas digital procedures represent phenomena through numbers or symbols. Most computers today are digital, but analog computers are still used.

attractor A pattern that 'attracts' a process. Fractals are the attractors of particular drawing algorithms. Equilibria are the point attractors for physical processes.

carbon cycle The sequence of steps that carbon cycles through an ecosystem. For example, plants fix carbon in their tissues, herbivores eat the plants, carnivores eat the herbivores, bacteria decompose the dead bodies of carnivores when they die, thereby passing the carbon into the soil so that plants can extract it to begin the cycle again.

cellular automaton (CA) An array in which each cell is an automaton (computer) with identical programming. Each cell interacts with neighbouring cells. The most famous example is John Conway's game of LIFE.

chaos Complex behaviour in dynamics systems. One characteristic is sensitivity to initial conditions. That is, small initial differences quickly lead to large differences in behaviour. Chaotic systems travel through many states without ever returning to their starting point.

codon A string of three DNA bases that code for a single amino acid.

computability The existence of an algorithm to solve a problem. See also **halting problem**.

cyberspace Thought space, the universe of information and ideas. Derives from Norbert Wiener's term 'cybernetics'. The concept of cyberspace can be compared to the philosopher Karl Popper's idea of a third world.

DNA/RNA (Deoxyribose or ribonucleic acid) The chemicals that store the genetic code.

DLA (diffusion limited aggregation) Formation of patterns or structures by aggregation in which the particles have to diffuse through a medium before sticking to a developing cluster.

disorder The term is used in a fairly precise way by physicists and engineers to mean a system with a lot of inherent randomness and therefore low predictability.

dissipative system A system that exchanges energy with its surroundings.

domain A logical section or grouping of sites on the

Internet, such as a country (e.g., '.uk' indicates Great Britain) or a type of organisation (e.g., '.com' for commercial sites).

emergence The appearance of large-scale properties in self-organising systems; e.g., seeing the wood for the trees.

entropy A measure of disorder in a system. A fundamental concept in thermodynamics and information theory.

environmental gradient A systematic change (across space or time) of an environmental property (e.g., soil moisture or temperature). An example is the drop in temperature noticeable as you climb a mountain.

fractal A pattern or process made up of many replicas at different scales. If all are the same it is said to be self-similarity, but if they are the same only on average, it is said to be self-affined.

graph Mathematicians have a precise meaning for a graph, somewhat different from everyday language. A graph is a set of points (nodes) some or all of which are connected by lines (edges).

halting problem The question of deciding whether a given Turing Machine program will ever finish. In general, the problem cannot be solved, that is, we cannot always tell whether or not a problem can be solved. A Turing Machine that will stop defines an algorithm for solving a problem.

hypercycle A cyclic pattern of chemical reactions in which each substance catalyses the formation of later substances. Hypercycles are thought to have played a part in the origin of life.

IP address The location of a computer. This is used by the Internet Protocol (IP) to pass messages to that computer.

iteration Repetition of a process again and again. Fractals are formed by iteration of the same process at ever smaller scales.

loop analysis This is a method of analysing the feedback loops in a complex dynamic system by tracing the signs of the interactions.

mapping A transformation that links input values to output values. For example, adding 1 to any number defines a mapping from each number X to X + 1; i.e., it maps the number 1 to 2, 2 to 3, and so on.

non-linearity The presence of non-linear terms or components (e.g., square, cubic) in a system or process. Non-linearity is essential for chaos. Interactions in physical systems are often non-linear.

phase-locking Constraining different processes to cycle in unison; e.g., clocks on a wall interact so that they all keep identical time.

phase space An imaginary space in which each dimension represents a variable or parameter of the system.

pixel A picture element. We can generate pictures as large chequerboards in which each square (pixel) is a different colour. Images on a computer screen are composed of thousands of pixels.

protocol Conventions for computers to exchange information over a network.

random variations The term derives from communications in which an orderly signal can be obscured by random background noise.

reductionist paradigm An approach to science in which we try to understand complicated phenomena by breaking them down into simpler parts. For example, to try to understand the human body we look at the functions of different organs and muscles.

stochastic Random values. For example, a stochastic process takes random values at any time.

synergetics The name Haken gave to phenomena in which lots of objects coordinate their behaviour and act in unison (e.g., a laser firing).

Turing Machine A hypothetical ultra-simple computer proposed by Alan Turing as a model for computability.

turtle geometry A method of drawing pictures by following the graph of an imaginary turtle around a plane invented by Seymour Papert.

UGC (unique giant component) A cluster of connected nodes in a graph that absorbs most of the nodes.

universal computation The ability of a computing system to carry out any computing task. It is often used in the sense of a Universal Turing, a Turing Machine that can simulate all other Turing Machines.

PREFACE

This book distils many years of practical experience with complexity. Our work on neural systems, environmental processes and many other questions in biology and computing has taught us the importance of interactions and other mechanisms that produce complexity. We want to share with you, the reader, some of the excitement we feel about this profound and fascinating new field of knowledge.

As scientists who apply computers to biology all the time, we have seen over the years the growing importance of the computer, both as a practical, working tool and as a metaphor for many aspects of the living world. Strange as it may seem, biology and computing are today very closely related disciplines. Not only is computing teaching us a lot about biology, computer scientists are also learning a lot from seeing how nature has dealt with many issues in computation and information processing. We try to draw out this close relationship again and again in different contexts throughout the book.

Complexity is an extremely rich area of research that draws together ideas, methods and phenomena from almost every traditional field of knowledge. Unlike recent popular books in this area, we do not dwell on the personalities of the scientists doing the research. Nor do we try to be encyclopaedic. Of necessity, there are many topics that we deal with only in passing, or not at all.

Instead, the main focus of the book is on what complexity is and how it impinges on our everyday lives. We use a lot of concrete examples—from starfish to travelling salesmen, from car crashes to the workings of the brain. Our hope is that our readers will go away with an appreciation of the main issues. If we help you to see the world in a new light, then we have succeeded. Finally, we invite readers who want to explore the world of complexity further to browse the on-line resources and information available through Complexity On-Line. This service can be located via our Web server at the Internet address http://life.csu.edu.au.

Terry Bossomaier
David Green
December 1997

Chapter 1
COPING WITH COMPLEXITY

The date was 5 May 1961. At Cape Canaveral, Florida, the United States' first manned space flight was about to get under way. Out on the launch pad, astronaut Alan B. Shepard was strapped into his Mercury capsule on top of a Redstone missile. This first flight was to be quick—an up-and-down, sub-orbital trip just beyond the edge of space.

Things were not going well. Time and time again engineers halted the countdown to check possible faults in one system after another. The problem was that everyone was dead scared of failure. After the Russians successfully sent Yuri Gagarin into space in April there had been tremendous pressure to get an American up there too. But American rockets had a history of blowing up in public. A similar disaster on this first manned flight would have seen American prestige and credibility blow up with the rocket.

As a result, every technician was frightened of fouling up. No one wanted to be the one who caused a disaster. So they checked and rechecked every minute detail of the systems under their control. The rocket could not be launched until every system had been confirmed as 'go'. But no one wanted to commit themselves in case they had missed some fault.

Although the flight was to last only fifteen minutes,

1

Shepard had been sealed in his capsule for over four hours. By this time his bodily functions began to take priority. The coffee he drank for breakfast had passed through his digestive system and wanted desperately to escape. No one knew what the effect would be of a sudden release of bodily fluids into his spacesuit. Shepard's body was taped with biosensors. Fluid could spark an electrical shortcircuit. A major concern was fire. Six years later three astronauts would be killed when a fire raged inside the pure oxygen environment of the first Apollo capsule.

Eventually he could hold on no longer. Instruments went crazy and a lot of biomedical data were lost. The go-ahead to launch was given before there were any more problems. At long last America's first manned space flight got under way.[1]

The point of this story is that a spaceship is an excellent example of a complex system. Thousands of separate systems have to work correctly if the spacecraft is to fly. With careful attention to every minute detail the risk of any one part failing is very small. However, there are literally millions of individual parts. So the overall odds of some part failing is depressingly high.

Another familiar example of complex systems is home electronics. Today electronic devices tend to be very reliable. Many TV sets, for instance, run trouble-free for years. This reliability is a result of advances in electronics. In the early days, TVs consisted of dozens of vacuum tubes. Like light bulbs, these valves tended to fail after a time, making visits by the TV repairman all too familiar. Unreliable vacuum tubes caused far more severe problems for the first computers. Whereas a TV set had perhaps a few dozen valves, early computers contained thousands. It took a large room full of valves and wires to build a device that was hardly more powerful than a pocket calculator. Failures were so frequent that valves were constantly being replaced. It was a challenge to make the machines work at all.

The story of computer programming is similar. Computer programs can be extremely complex. A major piece of software can contain many thousands of lines of code. In the early days of computing, the programmer would sit down and carefully plan out a program and draft the code on paper. This written version would then be typed up as a series of punched cards—one card per line of code. For a typical program these cards would fill a box. To run the program you would deposit the box at the computer centre and come back the next day to pick up the output. Inevitably the output would contain errors, so you would spend the afternoon poring over the code to debug the source and replace some of the cards before resubmitting the job yet again.

One rule of thumb in programming is that the number of bugs in a piece of new code increases exponentially with the number of lines. The shorter the segment of code the better your chances of getting it right. At first that doesn't make sense: the number of bugs in a few lines of code should be the same regardless of the size of the rest of the program. But what happens is that each new line can interact with existing lines of code in obscure and usually troublesome ways. Today this rule leads to the practice in which programmers write small additions to their code, then compile, test and debug all within minutes. In the early days, however, that approach meant months of painstaking development. In a later section we discuss a new trend in computing—a trend not only growing very fast but returning huge gains in productivity. The idea is crucial to managing complexity: encapsulating data with the procedures that can access and manipulate it.

Just as engineers developed procedures to reduce the risk of errors, so programmers have developed techniques

3

to minimise coding errors. Whereas engineers divide a complex machine into subsystems, programmers organise their programs into modules. Each module can be developed and tested before moving on to the next one. This approach really becomes powerful when you start reusing modules again and again, as building blocks. Its potential was first realised early in the nineteenth century by Ada Lovelace, the world's first programmer, who developed methods for Babbage's Analytic Engine.[2]

When the first modern computers were built after World War II, programs were all written in what is known as machine code. These were instructions that told the computer literally what to do. Instructions at this level are extremely detailed and convoluted. They are full of obscure commands like 'move the number in register 1 to register 2'. Perhaps the greatest advance in computing was to develop ways of giving names to groups of commands that are used again and again. This practice led to a special kind of program, called a compiler, that converted all those names of routines into the appropriate machine code. In this way programming languages were born.

Engineers adopt many strategies to avoid catastrophic failures. The most visible are monitors that report on operation of the system. We are all familiar with speedometers and fuel gauges in our cars. Complex machines such as airliners and spaceships need to have many times more sensors than a car has. The advantage of sensors is that problems can often be identified before they become catastrophic. However, each sensor is itself a new system that must also operate correctly. One of the most common problems in spaceflight is faulty indicator lights.

Another strategy is to build tolerances into parts and systems so that they can operate well beyond the range normally expected. A dramatic example of the dangers of poor tolerances was the Challenger disaster, in which a space shuttle blew up shortly after liftoff, killing the seven astronauts on board. The booster's O-ring seals failed because they became brittle when used in temperatures below those they were designed to tolerate.

Redundancy is yet another approach to coping with part failure. If a system is crucial you duplicate it. The space shuttle, for example, has three computers on board.

Without all of the above precautions, and many more, challenging activities, like flight or space travel, would be impossible. And yet, even after all those precautions, perhaps the biggest problem of all remains. And it is that the interactions of different systems with each other are the greatest source of complexity. Just look under the bonnet of your car. The packing of all the complex parts under the hood is a triumph of design. Change anything and that change affects everything else. Enlarge the carburettor and the exhaust hose no longer fits. Move the exhaust hose and it could overheat the electrical wiring.

Yet, despite all these techniques and precautions, many of us have had the experience of staring in frustration at the computer screen after something has gone wrong. Maybe the machine has just frozen, maybe a file you've been working on for the last hour has disappeared—whatever it is, it's usually annoying! We don't have a clue what has happened, and we don't need to be told that computers are complex. Sometimes the behaviour is really spooky. Installing a new software package causes tried and trusted programs to play up. Yet today's digital computer is quite a simple rule-based device, with nothing like the sophistication of a biological brain. These screen lock-ups or other frustrations arise usually through failures in a program or sometimes in the interaction between programs, perhaps in the same computer, perhaps over the Internet.

The possible complications of interactions between systems are endless, and often unpredictable. That is why a completely unforeseen situation, such as Shepard's fall from grace, is always on the cards in spaceflight. It is also the biggest headache for designers of any large system. The packing of a car's engine under the bonnet did not just happen overnight. It is the end product of literally decades of gradual improvements and design changes.

FROM CATS TO THE WEB OF COMPLEXITY

It's as difficult as herding cats, was how a frustrated dean once described running a university faculty. Cats are independent creatures and hard to control. Getting lots of them to be well behaved is impossible. But you don't have to be as complex or independent an entity as a cat to get an unpredictable system. Isaac Newton used to complain that thinking about the moon made his head ache. The reason? Try as he might he just couldn't solve the equations of motion for the system of the earth, moon and sun. What he did not know (it was not proved until several centuries later) was that his efforts were doomed. No solution exists for the so-called *three body problem*, of which this is a special case.

Nor do you need many cats for chaos, either, as a kitten with a ball of wool will soon demonstrate! As the kitten chases the ball around the house it unravels, leaving chairs, tables, just about everything, tied together. Each time the ball hits an obstacle it bounces off at an odd angle, since it is not a perfect sphere. Unless we can find an end to the wool, we have no idea where the path started or stopped. Here we have one agent, following a complicated path, going round and round the same area along slightly different paths each time.

6

We might call this the *iterative* approach to complexity, the repetition of simple acts again and again.

A spider's web is another example of a complex pattern formed by a single agent. The spider traces a complicated path in time as it spins: the pattern of the resulting web in space is a record of the spider's path. We can see here a very important idea which we will meet frequently. Complexity is seen *not at the level of an individual,* be it animal, plant or mineral, *but in the larger picture,* which may be an aggregate view over time or space. In the case of the spider's web we can see a very clear pattern; in the case of the ball of wool there is no obvious pattern at all. But patterns, intertwined paths, structures of all kinds, are hallmarks of complexity.

So just what is complexity?

In essence, complexity is about the ways in which the world is put together. As the above examples already show, when you put lots of things together, such as subsystems in an airliner, the interactions between them can be extremely complex indeed. It is these interactions that turn something from being merely complicated (having many elements) into something truly complex.

If you roll a billiard ball across a billiard table, then any good billiards player can predict the path that it will follow. If you roll two balls, then the problem is still simply a matter of calculating each ball's path individually. However, there is also a chance that the two balls will bump into each other. Add a few more balls and the problem of keeping track of paths and collisions becomes very difficult and predicting the state of all the balls becomes well nigh impossible.

Suppose now that you roll (say) 100 balls around on the table. With so many balls, they will be bumping into each other all the time. The problem of computing individual paths becomes virtually impossible. Such a system is essentially unpredictable at the individual level. It is hard to record, in the sense of logging where everything is, and devoid of simple patterns. However, if what we want to know is not the path of any particular billiard ball but the appearance of the entire table, then with large numbers of balls the problem starts to become simpler again. We no longer need to trace every ball exactly. Instead we can look at average behaviour. The myriad individual interactions average out and we can make sensible predictions about average speed, average time between collisions and the average distribution of balls on the table.

Here we have an example of a key idea in complexity: *emergence*. The average properties described above *emerge* out of countless individual interactions, which in a sense cancel each other out. The billiard ball model is the starting point for the kinetic theory of gases. For example, in a container full of very many particles (atoms, molecules, whatever) whizzing around at all sorts of speeds, bumping into one another and into the walls of the container, we have one single property that is stable with respect to type of gas, shape of the container, nature of the walls and so on. That property is temperature. We call it an emergent property because we wouldn't have expected it just from looking at individual particles and we can't measure it for any single particle. Unfortunately emergence is almost as hard to define as complexity. Emergence is one of the hallmarks of complexity. Out of a set of simple processes comes something unexpected on a global scale. We shall see many examples throughout the course of this book. In the end we shall have a feel for what it is, but still perhaps not a formal definition.

Temperature is a good example of an emergent property. What is it? We can measure it on a macroscale and we have all sorts of physics dependent upon it. But at a molecular scale it disappears. We have only many molecules or other particles (nuclei in a plasma, etc.) with a wide variety of velocities (and consequently energies). It is only the collective distribution that gives rise to the single scalar quantity temperature for the assembly.[3]

What is true of billiard balls is true of other systems too. Whether it be gas molecules in a jar, stars in a galaxy, cells in our bodies or people living in a town, it is the interactions that make each system complex. Traditionally each of these systems has been studied separately. Gases have been in the domain of thermodynamics, cells in biology, and towns in sociology. Only recently have scientists begun to appreciate that, despite their obvious differences, all these systems share important properties in common. Is the whole greater than the sum of its parts?

A PARADIGM SHIFT IN SCIENCE?

As the science of putting things together, the study of complexity is a radical departure from the traditional way in which science has been done. Traditional science uses what is known as the *reductionist paradigm*. The idea is that you can understand the world by breaking things down into their components. Complexity is exciting because, while the components may be trivial, their *interaction* often generates wild and unpredictable behaviour.

For instance, if you want to understand the relationship between a forest and the environment, then you look at plant physiology. You study the way plants pump water and you study the biochemical pathways involved in respiration and photosynthesis. You perform experiments in which you measure the exact rates of water

9

transpiration under different temperatures and air pressures. This is the reductionist paradigm. It is a way of doing science that has served us well for hundreds of years. It has led to countless discoveries and has led to the technical wonders that we are so used to today.

So, at the end of all your experiments on plant physiology, you know exactly how a forest works. Or do you? You certainly know a lot about the workings of an individual plant, and that can tell you a lot about the nature of forests. It tells you where plants can grow and where they cannot. But does it tell you everything?

Fortunately for field ecologists, the answer is a resounding no. For the truth is that a lot of features in a forest are the results of interactions. Studying an individual plant does not tell you how it is going to interact with other plants. Trees in a forest may seem stately and quiet. But in reality a forest is seething with competitive and symbiotic interactions. They have a telling effect. Plants compete with one another for space, for light, for water and for nutrients. They disperse seeds. They provide homes for birds and animals. They are attacked by insects and other herbivores. They create leaf litter, which develops into soils and provides an environment for millions of insects and micro-organisms. Without all the richness that these interactions bring, a forest would be little more than a sea of potential telegraph poles.

The reductionist approach is to reduce everything to the absolute minimal level. So if we could understand a forest by studying plant physiology, why not understand plants by studying their cells? Why not understand cells by studying biochemistry? Why not understand biochemistry by studying nuclear physics? The ultimate expression of this approach is the quest for a unified field theory. Einstein spent the last years of his life searching for a way to link gravity and the electromagnetic field with the strong and weak nuclear forces. The quest continues today, with many successes and failures along the way.

But if the quest for a unified field theory ultimately succeeds, if someone does create a 'theory of everything', will that be the end of science? Should we all pack up our bags and go home? This is a far from trivial question. Paul Dirac, predictor of anti-matter, leading quantum theorist and Nobel laureate, asserted that chemistry had become a branch of pure mathematics! Can we start from these basic rules of fundamental science and predict how a large system will behave? For a long time the answer seemed to be yes, prompting all sorts of arguments about the existence of free will and other matters of metaphysics. We now know that it is not always true. Even using huge computers? In fact the amount of computation required grows so rapidly that, for even simple chaotic systems, prediction on large time scales is just not feasible, ever.[4] There are other complications to do with quantum mechanics, but we will not dwell on them here.

Simulation of a complex system may be very useful practically. Boeing design aircraft on computers rather than using wind-tunnel models, on grounds of speed and cost. But simulation doesn't help us *conceptually*, in understanding the rules of behaviour at the higher level. So are there methods to develop that give us the rules of the emergent system? Just as with our attempts to understand the nature of a forest, there are many things that the lower-level phenomena cannot explain. In fact there is a subtle computational argument for believing this to be the case, which we shall consider in our discussion of the fundamentals of computation.

The most intriguing example of an emergent phenomenon is surely human consciousness. The human brain has many nerve cells, but the number of individual types is very limited, certainly less than 100 but not much more than ten. The dynamics of a neuron are no deep mystery and the way neurons adapt (learn) is still not completely solved, but the broad outline seems to be there. So where or what is the human mind? This deepest of philosophical questions is still with us and not obviously soluble through ever deeper study of the behaviour of nerve cells. It is somehow a phenomenon which emerges from the grand ensemble.

To truly understand the world it is not enough to take it apart. We also have to learn how to put it back together. The reductionist approach has been an enormously powerful tool for science, and it will continue to be so. However, we cannot blind ourselves to the need to understand just how and why the whole is so often greater than the sum of its parts. Nor is this need merely curiosity. The ultimate complex system that we have to deal with is the world we live in. If humankind is to survive and prosper then we have to learn how to manage this complex system of which we are a part.

CHAPTER 2
COMPUTATION

Before we examine the links and analogies between science and computation we need to ask what computation really is. Is it just what computers do? Hardly, since computers vary dramatically from month to month and we can do most things with a pencil and paper (maybe not so fast!) that a computer can do.

The computers we use in our homes and offices are usually derived from the model put forward by John von Neumann. In this model computers consist of a memory and a processing unit. The processing unit fetches data from the memory, carries out arithmetical or logical operations and writes new data back. The instructions for the processing unit constitute the program. Additional operations include control operations which allow the processing unit to repeat, or iterate, a set of operations or to select different instructions according to some data value.

This is a very functional description. At a more abstract level what our computer does is to map one set of numbers (integers) to another. We say that the output is a function of the input; alternatively, a *mapping* from the input to the output. For finite data sets for input and output we can make a list of which output goes with which input. This could get very cumbersome, but it sets an upper limit to the size of a program and there is no reason why we shouldn't be able to compile this list for any given case. However, when the range of possibilities becomes infinite, encompassing all the integers, it turns out that not all such mappings are computable. We could say that 'computable' just means 'can be computed on a PC or a Mac'. But computers come and go very quickly, so what we need is not an example of a present-day computer but some more abstract general model of a computer. Enter the Turing Machine.

TURING MACHINES: THE ESSENCE OF COMPUTATION

Alan Turing proposed a definition of 'computable' based on a simple abstract computer which bears his name and is now the building block of elementary courses in computer science. The Turing Machine (or TM for short) is a model of a computer reduced to the absolute basics. In Turing's model the computer consists of two elements: a tape and a controller. The tape contains three sorts of symbols, blanks, zeros or ones. These symbols can represent information in various ways. For instance, a string of six ones might represent the number 6. The controller is a device that performs various operations on the tape. It can move the tape one step backward or forward and it can read or write one symbol at a time. To make it work the controller has two other features. The first is its program, which determines its behaviour. The second is its 'state', which is somewhat like a person's mood in that it affects the way the controller behaves.

So, given a tape with some symbols on it and a controller with a number of states, the TM will go into action. The program consists of rules that tell the machine what do to for each combination of state and

14

tape symbol that it encounters. The following list sum-
marises the rules that make up a simple program.

STATE	TAPE	ACTION
0	0	advance the tape forward
0	1	change state to 1, advance the tape forward
1	1	advance the tape forward
1	0	write 1, change state to 0, advance the tape forward

Suppose that the TM starts in state 0 and that we
feed into it a tape with the following string of symbols
on it: 00011110000. As the machine reads this tape it
behaves as shown below.

INITIAL STATE	INPUT SYMBOL	OUTPUT SYMBOL	FINAL STATE
0	0	0	0
0	0	0	0
0	0	0	0
0	1	1	1
1	1	1	1
1	1	1	1
1	1	1	1
1	0	1	0
0	0	0	0
0	0	0	0
0	0	0	0

The net result is that the machine has increased the
number of 1's on the tape by one. That is, it has changed
the number 4 to 5. So the function of this machine is
to add one to any number fed to it. Now this may not
exactly be a very profound result, but this is just a simple
example. The point is that it *is* a well-defined function.
What Turing showed was that, simple though it may be,

15

Figure 2.1 The operation of a Universal Turing Machine. One tape holds data, the other a program.

DATA

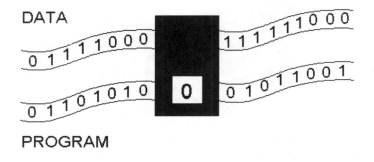

PROGRAM

we can design machines like this to carry out *any* arithmetic computable function we can think of.

An important limitation of the TM is that each machine performs one, and only one, function. To cope with this problem Turing introduced a second tape drive. This new device, called a Universal Turing Machine (or UTM for short), reads its program from the second tape. Turing showed that the UTM is capable of emulating any simple TM. This property, called *universal computation*, is the essence of any modern computer.

Now, as a device the UTM is hopelessly impractical. To try to solve real problems with it would be extremely slow and awkward. Nevertheless it does encapsulate the main features of any modern computer. The UTM provides a useful model for asking general questions about computing.

Before this *digital* model of computing, based around mapping sets of integers onto each other, people talked of *analogue* computing. Lots of physical devices, such as thermostats, are analogue computers. The old black vinyl records are an analogue recording medium, unlike CDs

16

which are digital. On a vinyl record the sound signal is related directly to the changes in shape of the record groove, the general principle of analogue systems. Some hi-fi enthusiasts maintain that analogue records are better than digital, but, if that were true in the early days of digital, the difference is fast disappearing. One of the beautiful features of the Turing model of computation is that *everything* in the classical world is computable digitally.

THE HALTING PROBLEM

An important question in computing is whether particular problems can be solved. One way to approach this question is to ask how long it will take. If a TM can solve a particular problem, it will stop when it has finished. It will stop after a finite number of steps.

But what if the TM cannot solve the problem? Will it ever stop? This question—called the *halting problem*—turns out to be unanswerable for any arbitrary controller. There is no function that a Turing Machine can represent that will determine whether an arbitrary Turing machine program will terminate. But any function represented by a finite controller that does stop is defined as a computable function. The unpredictability of emergence arises from the impossibility of determining if a Turing Machine will halt.

Turing's machine is just one possible procedure for carrying out a functional mapping: could there be better ones? Lengthy research over the subsequent decades has not found an alternative. In fact, Alonzo Church, another prominent figure in the theory of computation and Turing's PhD supervisor at Princeton, asserted that *effective calculability* is independent of formalisation, a thesis which still stands today.

In modelling science by mathematics, we implicitly

17

assume that changes in state in the natural world are all computable. So far, there is no counter example, or likely counter example. Yet it is still possible to find people with an intuitive (only) feeling that there are things a brain does which do indeed lie outside the class of computable functions. But the brain is steadily yielding its secrets. We now know much more about biological neural systems and can also model neural computation quite successfully on computers. From the classical standpoint we see no reason to doubt that the computational strategies of the brain will eventually be fully understood.

Computational hierarchies

Complicated computer programs are difficult to write and, since the early days of the development of programming languages, modular structure has been a key tool in managing complexity. The next step beyond this was to encapsulate both data and program into independent modules, or objects. This new paradigm swept through computer programming during the 1980s. As computers have entered more and more of our lives, as hardware improves rapidly, the need for robust software gets ever greater. Yet the design difficulties have proved so great that people talk regularly of a software crisis. Many see object-oriented methods as a major step forward. An object is a collection of data and the operations that access and manipulate that data. By hiding the data, and allowing access to it only through a limited set of special methods, the complexity of software development can be broken down into manageable and provably viable pieces.

Natural complex systems exhibit modularity almost everywhere, from the building blocks of evolution to the structure of the human brain. Thus the object-oriented approach has the right feel about it. But alongside the

18

development of object-oriented methods the networking of computers has mushroomed, bringing with it the need for objects to be accessible across networks and for objects to be self-describing. Soon we will have the option where one object may go searching on computer networks for other objects about which it knows nothing in advance. The stage is set for emergent computation on computer networks! We do not yet know what will happen.

A PHOENIX RISING FROM THE ASHES

'We are on the threshold of a new era in computer architecture.' So began an excellent computing text, *Highly Parallel Computing*, by Almasi and Gottlieb, published in 1989. At that time, with defence money pouring in, massively parallel computers were the cutting edge of supercomputing. Machines such as the *Touchstone Delta* (Intel) at Caltech or the *Connection Machines* produced out of Boston by Thinking Machines Corporation vied with each other for supremacy in a range of benchmark mathematical, scientific and engineering problems. Other companies such as INMOS and Texas Instruments were building plug and play parallel computing chips. Parallel computing was the way forward, or so it seemed.

The obvious advantage of parallel computers is that they can perform many tasks at the same time. Imagine a supermarket checkout during peak hour shopping. A shop with just one checkout is like a computer with a single processor. During busy periods such a shop is liable to have long queues of waiting customers. Supermarkets solve the problem by adding more checkouts. As a group, twenty checkouts can process customers twenty times as fast as one can. Likewise, a computer such as the Connection Machine CM2 with (say) 65 536 processors can carry out a lot of similar calculations thousands of times faster than a single processor.

But parallel computers turned out to be hard to program. Getting messages between processors and making sure that all processors are fully occupied all of the time is far from easy. It's not only difficult, it's difficult to automate, making parallel programming a rather specialist art. Worse still, parallel computers had to survive in an extremely competitive computing market. By 1996 the situation had changed radically from 1989. Almost every computer manufacturer dependent on parallel computing was in trouble: Thinking Machines Corporation had been dismembered after serious financial difficulties; Cray Research had collapsed; Intel had ceased work on its range of parallel computers; Convex went under. (Intel of course had a booming semiconductor business.) Two things had happened: firstly, the collapse of the Iron Curtain dramatically cut defence funding and the Californian freeways once full of cars into Silicon Valley were almost overnight congested no longer; secondly, the growth in microprocessor performance was greater than anybody had expected. Make no mistake, the most powerful machines on the planet are still parallel: but the mass market opportunities are gone. In their place we have distributed computing with independent machines loosely coupled by computer networks.

One method of distributing computation is to carve up a large problem into many small, separate cases and to assign each case to a different machine. Computer networks such as the Internet serve as a parallel computer, but one in which communication between the processors is very poor. For instance, simulations of the world's weather divide the earth's surface up into many cells and assign each cell to a separate processor. However, to simulate the circulation of wind and water, those cells have to exchange information with neighbouring cells constantly.

Large machines with many interconnected processors on special high speed communication frameworks are much better suited to problems where the different cases have to exchange information, and problems such as this are still best handled by the big (very expensive) purpose-built parallel machines. The leading edge of high performance computing is still parallel.

Other problems do not require this tight coupling, and heterogeneous computing networks are viable. Updating all the records in a large database, for example, would just involve each processor taking its share of records. Communication between the processors would not be necessary.

COMPUTATIONAL COMPLEXITY

We now have a simple model for universal computation. Next we need to look at what constitutes complexity for a computer and see how this might relate to complex systems in the natural world.

Hard problems

In Douglas Adams' novel *The Hitchhiker's Guide to the Galaxy,* scientists ask the computer Deep Thought whether it can find the solution to the ultimate question of life, the universe and everything. The computer replies that yes, it can, but the solution will take some time—eight million years. As outrageous as this period may seem, it is not difficult to find problems that would take much longer than eight million years to solve. The question of time to solve a problem is relevant whenever we try to scale up from a small problem to a large one. Most problems can be solved quite easily for small cases, but what happens as we increase the number and complexity of cases involved?

What *makes* a problem hard to solve on a computer? The complexity of a computational problem has been thought of as being composed of the time it would take to solve the problem and the working space needed to store all the intermediate results. In computer science the first of these, the time taken as a problem increases in size, often receives the most attention. We could use some measure of the number of steps it takes to solve a problem as a measure of the time it would take. Since computers are getting faster all the time, it would make little sense to use an actual time, although in principle we might standardise on some particular machine. But then we would have to rely on standards museums to keep such machines alive and working, which does not seem a very practical idea. Instead, what we do is to determine how the number of steps grows in size as the size of the problem measured in some way increases. The time taken or the number of steps required depends of course on how we go about the solution—the *algorithm* we adopt. There are exact and approximate algorithms and also good and bad ones.

One of the lessons from natural computation (discussed later) is the emergence of fast approximate solutions. Needless to say, these are of potentially great interest in many application areas. To make this a little easier to grasp, let's look at a couple of examples. An easy problem is a simple search. Suppose we have a number of identical containers in which we have to search for something. If it takes the same time to search each container, then the time taken will increase linearly with the number of containers: if we double the number of containers it will take twice as long and so on.

A somewhat harder problem is to determine which two of a set of cities are the closest together. To create a table of distances between cities we need to do a calculation for every possible pair. For each new city that we add to the list, we have to calculate its distance from every other city. For 5 cities, we need to calculate the distance from the first city to all 4 others, the distance from the second city to the 3 remaining cities, and so on. So the total number of calculations is $4 + 3 + 2 + 1$, giving a total of 10 cases to calculate. For 100 cities this number grows to 4500 cases. Any home computer could still knock this over in no time flat. In fact the size of the problem increases with the square of the number of cities. We say that the growth is *polynomial* in the number of cities.[5]

A well-known hard problem is the travelling salesman (this should now be travelling salesperson, but we need to stick with historical usage). Suppose that our salesman needs to visit each of a set of cities. Then, to save time and money, he needs to know the shortest route that visits each of the cities just once. Unlike the simple table, if we now add a new city to our list then we have to check its effect, not on each other city, but on every possible tour of those cities. So each new city multiplies the number of possible routes. For 5 cities there are only 12 cases to examine (the length of the path is independent of the starting point and direction). However, the size of the problem escalates at an alarming rate. The figures in Table 2.1 show just how rapidly the complexity of the problem goes off the scale.

Suppose that we have a supercomputer capable of checking a billion cases per second. Then just 28 cities are enough for our supercomputer to take longer than the age of the universe to solve the problem. This is not reasonable at all! In fact, the number of routes grows *exponentially* with the number of cities.

Computer scientists make a precise distinction between classes of problems on the basis of the rate at

Table 2.1 Number of tours as a function of number of cities

Cities	Tours
3	1
4	3
5	12
6	60
7	360
8	2520
9	20 160
10	181 440
15	4 358 914
20	10 000 000 000 000

which growth occurs. The Travelling Salesman Problem (TSP for short) belongs to a class of problems known as 'NP complete' (NP is short for 'Non-deterministic Polynomial'). These are problems that quickly grow unmanageable as their size increases. In many of these problems the complexity stems from permutations and combinations of simple elements. In the TSP, for instance, each path can be written as a string of cities ABCD . . . or whatever. So the number of cases depends on the number of ways in which we can order a string of symbols.

This abstract description reflects an important feature of computer science: many apparently different problems are equivalent to one another. So the solution to one problem typically allows us to solve many related problems. For instance, some similar problems include trying to find the shortest route through a maze of streets or trying to use the least amount of wire when installing electricity in a house. At a more general level, the TSP is one special case of the so-called scheduling problem. Other examples of scheduling include arranging classes in a school, timing traffic lights, land-use planning, and programming work on machines in a factory.

SEARCHING AND SOLVING IN THE
NATURAL WORLD

Animals spend most of their lives searching for food, mates and shelter, but searching is also used as a computational metaphor, where we think of searching in a solution space. Unlike the TSP, search problems are often solvable in a reasonable time. First of all, we do not need to test every possibility to solve the problem. We can stop as soon as a solution is found. Secondly, we can often discard many cases without checking them individually. For instance, suppose that we are searching a medical database for people satisfying a set of selection criteria. Then if a person fails the first criterion we can immediately discard that record; there is no need to check all the other criteria. So what do we do about problems like the TSP that need millions of years to solve? City planners simply cannot wait around for centuries to solve the problems of choked urban roadways. And to say that a problem is insoluble helps no one. Real world problems need real world solutions, fast, and roughly but not exactly correct.

We can begin to get somewhere if we note two points. Firstly, people usually need a good answer, but not necessarily the best. Secondly, as in life it pays to compute smarter, not harder. Earlier arguments about the TSP concern 'brute-force' methods in which the computer laboriously checks all possibilities, one after another. Brute-force methods are not always feasible. How *do* travelling salesmen plan their routes? One approach is to use a technique that might be termed 'divide and rule'. One practical aspect of travelling is that you have to stay somewhere. It is more convenient to stay several nights in one hotel than to be constantly moving. So travellers usually base themselves in major towns and make a series of short side trips to nearby towns. So if you had to visit 20 towns you might divide them into (say) five groups of four towns each, with a central base in each group.

The problem of planning the shortest route thus reduces to finding the shortest route around the five base towns. So instead of a vast number of cases the traveller has only twelve cases to consider—a task that can easily be done by hand on the back of an envelope. The basic idea embodied in the divide and rule approach is to break a big problem down into small ones for which a brute-force approach is viable. This is yet another example of a computational hierarchy.

One way of visualising the business of solving complex problems is to think of a *solution* landscape in which all possible solutions to a problem are laid out on a map. Solutions that lie near one another are similar in some respect. Also, if we suppose that elevation represents the quality of a solution (e.g. the travelling time in the TSP) then we can imagine solutions as looking like a landscape with hills and valleys. So the task of examining one potential solution after another is like wandering around in the landscape where the desired solution is the highest hill (or deepest valley, depending on the nature of the problem). This image of a solution landscape derives from problems in which solutions can be expressed as numeric values of parameters. For instance, efficient production in a factory might depend on a number of factors such as length of production runs, stores of raw material and so on. If we treat the values of these parameters as akin to latitude and longitude then we can literally plot values in a solution space.

Supposing that we can structure our search as a solution space, how does that help us? Well, first of all it helps us to reduce the search time. For instance, if we start anywhere in the landscape we can immediately find a better solution by wandering about. How does a blind mountaineer find the top of a mountain? The answer of course is by going uphill. So, starting any-where, we simply go uphill until we cannot go any higher. For simple problems this approach (called 'gra-dient ascent' or 'hill-climbing') is a good way to find the best solution.

In general, however, a solution space may be a complex landscape containing many hills and valleys. The problem is that hill-climbing yields only a local optimum. There is a risk of becoming trapped in the foothills and missing the mountain altogether. What we need is a method of surveying all the major hills in the landscape.

This is an area where computing can learn from nature. Nature has evolved many ways to solve complex problems. These have provided inspiration for many algorithms, or methods for solving problems. Take an ant colony. The colony needs food to survive and has (literally) to search the landscape to find food. As we all know, the colony sends out scouts to locate food. Note that it sends out many scouts, not just one. The scouts have no knowledge of the landscape, nor do they know which areas are more likely to hold food than others. They have no intelligence, so they cannot plan a systematic search to take in every corner of the local area. So what do they do? Simple. They just wander around at random until they find something. That something may be food, or it may be some clue, such as a scent, or a pheromone trail left by other ants.

When they do find something the ants head off to the nest, leaving a trail so that other ants can follow the same route back to the prize. This strategy is far from perfect. Ants do miss a lot. They often walk straight past food without detecting it. However, that does not matter. The goal is not to find every scrap of food but to find enough food. What matters is retrieving the food that they do find quickly and efficiently before someone else does.

Every species has its own foraging strategy. Contrast the ants with (say) an eagle. Unlike an ant the eagle literally does have an overview of its entire environment. However, its food, such as mice, rabbits or lizards, tends to be rare and tries to evade detection. So the eagle's strategy is to hover in the sky scanning the ground for any hint of movement. The chances of finding a rabbit under any one bush is small. But by scanning the whole landscape the eagle is able to spot any error made by any rabbit in the entire area.

Evolution itself can be seen as a search technique. One way of looking at evolution is that it is a grand experiment—a search in which nature is constantly experimenting with new designs for making living things. Good solutions are animals that survive and breed to produce offspring. Two things to note about evolution and natural selection are, first, that it throws up not one solution but lots of solutions and, second, that its starting point is always a previous solution that was successful. This appreciation of evolution led John Holland to develop a computational version of evolution. This technique—the genetic algorithm—mimics the above process to solve complex problems.

Evolution, brains (neural systems) and the immune system have all spawned computational methods. We consider some of these methods now.

SHORT PROGRAMS FOR ORDERED SYSTEMS

Staring into the fire, in the days before television, we would see the flames shape themselves into familiar objects such as faces. This idea of patterns in random systems appears in many places: there is even a psychological test based upon what people perceive in ink blots. This quirk of human nature conceals two very important ideas. The first is the nature of order and randomness. The second is the search for order which drives our brains.

If you look at an active television not tuned to a station you see a random pattern, so-called white noise. Now, just as in the fire or an ink blot we might be able to interpret this rapidly changing pattern as being like some natural image; in practice we don't see anything really approaching the natural world. If we think of the television picture as being made up of an array of *pixels*, each pixel having a particular colour, then from the number of pixels and number of colours we can calculate the maximum possible number of television pictures.[6] It's not difficult to imagine that we would never in our lifetime see a real picture and thus we can accept intuitively that the proportion of random pictures is very high. In fact randomness is a tricky concept and it takes some sophisticated mathematics to make it precise.

We have seen the notion of the time it takes to solve a problem as a measure of complexity. This is a fairly old, traditional notion. However, in the 1960s the great Russian mathematician Kolmogorov, founder of modern statistical theory, came up with another complexity idea. As with many breakthroughs in science more than one person may be very close, and in fact what is now called Kolmogorov complexity was essentially invented simultaneously by Chaitin in the United States, and Solomonoff, also in Russia.

The idea is that any object can be described by a computer program. The length of that program is the complexity of the object. So if we compare a picture of a human face with a chequerboard pattern, the face is much the more complex: the chequerboard pattern can be described by a very simple formula. (We might compare abstract impressionist paintings along these lines: Mark Rothko's vast canvases have a few more or less uniform colours in large patches; Jackson Pollock's equally vast canvases are extremely complex with much tiny detail, and are *very* difficult to describe.) It might seem at first sight that this would be a very uncertain quantity, strongly dependent upon the computer and the language used to program it. Fortunately this is not the case, when the programs are large enough. We can see this simply by taking some common language for the description and just adding a translator at the beginning for any other language. The translator will be of a finite size and fixed independently of the problem.

Many complex systems display elaborate and complicated surface phenomena, although beneath the surface very simple rules are at work. Thus they may be modelled with a short program. At first Kolmogorov complexity seemed like a route to the measurement of complexity itself, but since the most random objects have the highest values it does not fully capture the notion of complexity. Murray Gell-Mann, Nobel Laureate and co-founder of the Santa Fe Institute, expresses it neatly: surface complexity with underlying simplicity. As a measure of complexity we need a combination of short program and complex output.

Noise, entropy and information

Putting names to faces can often be difficult: not long ago, at a conference on information theory at Brighton in England, somebody casually registered under the name of Claude Shannon. The conference got under way without anyone taking particular notice of this registrant. Then whispers started that this was in fact *the* Claude Shannon, who wrote the textbook for communication theory and demonstrated that exact communication was possible even in the presence of *noise*. (Not all pioneers are attention seeking!) Although in principle we might measure something with arbitrary accuracy, in practice noise prevents us from doing so. Shannon, however, showed that it was possible to define the information capacity of a channel, even if this channel was noisy. Furthermore, the channel could transmit information encoded in such a way that, provided this capacity was not exceeded, the signal could be reconstructed perfectly, at the other end.

Thus for present purposes we can just ignore the arbitrary accuracy of physical measurements and just assume that they are discrete, with the interval between gradations being determined by noise.[7] But in fact there is a more fundamental result, established only in the early 1980s, which relates computation directly to energy.

Thermodynamic cost of computation

For some time, debate has raged over whether or not computation requires energy. True, the Turing Machine is a hypothetical machine, with zero weight tapes and frictionless movement. So just how far could we push the limits of computation as computers get better and better? Is there some inevitable loss of energy as heat, as with all machines? Perhaps not: Landauer and Bennett at IBM established that computation in terms of changing information from one form to another does not require any energy, and we should be able to build a perfect computer.

But the *destruction* of information takes energy. In very simple terms, if we carry out a very complex calculation but save all the intermediate steps, then we can carry out the calculation backwards: it is reversible. But as soon as we destroy the intermediate working, then reversibility is lost and at this point energy is expended.

Traditional statistical thermodynamics has always recognised the intricate relationship between order and energy. Breaking up ordered systems, as in the change from a solid to a gas, involves an increase in *entropy*, or disorder. Thus going from solid to gas requires less energy than that required simply to break the bonds holding the solid together. These ideas go back to Gibbs in the nineteenth century. But a much newer result is that the destruction of algorithmic information also has a penalty.

In what sense can we think of the world as computation? The Turing model focuses on inputs and outputs, but a physical or biological system is analogue. Even if we ignore quantum mechanics, there is still a difficulty in moving from simulating physics on computers to physics being computation, at least in the Turing sense. Some biological processes are obviously rule-based and these map readily to the Turing Machine model. But an alternative is to think not in terms of algorithms but in terms of the creation and destruction of information. Now we can put a precise limit on the cost of destroying a single bit of information, kT, where k is Boltzmann's constant and T the absolute temperature, exactly the same energy as per degree of freedom of a molecule in a gas.

How efficient, then, are today's computers? So far they are nowhere near the above theoretical limits. Recent work at the University of Cambridge has estimated precisely the energy used in neural computation in insects. It too is a factor of millions above theory. Biological sensory systems have evolved to get very close indeed to the physical limits: many animals can detect single photons and some have hearing that is limited only by thermal noise in the membranes of the ear. No creature in the universe could do better. Computers, it seems, still have a long way to go.

And so what of quantum mechanics?

The complexity of the world of classical physics is rich enough. But then we come to quantum mechanics, which replaced Newtonian mechanics at the beginning of the twentieth century. David Deutsch, a physicist at the University of Oxford, created a quantum analogue of the Turing Machine. It can perform some calculations faster than a classical machine can. Unlike a conventional Turing machine, the Deutsch Machine has an array of tapes that are blended when the machine stops. Theoretically this is a very interesting field. There is active research in how to build a quantum computer and research into the sorts of problems that might be solved faster by it.

David Deutsch has shown that the computation of the real world needs a new type of quantum Turing Machine, while Roger Penrose has suggested that quantum phenomena may be fundamental to the working of the human brain. Although these ideas are profoundly original, there are still many classical avenues to explore before we give up on the mysteries of the human mind.

Later we will talk about local rules in cellular automata, neural networks and self-organising systems and how local rules produce interesting global emergent phenomena. But the message of quantum mechanics is that the world really is non-local. In the next section we describe the remarkable experiments that demonstrate fundamentally non-local phenomena. Yet we do not know if these have any macroscopic effects.

Non-locality

The two twentieth-century revolutions in physics—relativity and quantum mechanics—have in common some very unintuitive phenomena. In quantum mechanics one of the strangest is *non-locality*, a sort of instantaneous action at a distance. Very precise experiments have tested this notion. Quantum mechanics gives us two complementary perspectives on light. For example, we can view it as a wave motion or as a stream of particles called photons. Each photon has a spin, just like a top or the rotation of the ear. For complicated reasons, we can't give a very precise value to the size of a photon. Suffice it to say that photons are *very* tiny—much smaller than the smallest atom. Yet the most recent experiments involve matching spins of single photons over a distance of several metres! (See page 35.)

There isn't any causal link, though, since this would violate the principles of relativity, where no communication can occur faster than the speed of light. Since effects such as this are so unlike anything that we encounter in our daily lives, that we encounter with our senses, we resort to stories and analogies to picture what is going on.

To understand non-locality, let us consider a simple story, which follows Roger Penrose's Quintessential Trinket argument. Imagine a gift shop situated in a remote part of the galaxy. It is important that it be a long way away, since we want to be sure that no secret messages, travelling at the speed of light, could arrive in a short space of time. One of its gifts is a little goblet set packed in pairs, one silver, one gold. The shop will dispatch one of each pair to one of any pair of friends, pre-wrapped so that even the people in the shop don't know who gets the gold and who the silver, except that there is one of each. So if one goblet is sent to you and the other to a friend light years away from you, it will not be until you open it that you know which each of you got.

There is nothing difficult about this: the goblet in *your* package is either silver or gold, but which it is just not known until the package is opened. The same goes for your friend.

But there is another gift for each of you: a Quintessential Trinket, a little toy in the form of a dodecahedron (a solid object made up of twelve faces each with five sides). On each vertex is a button: pressing the button may do nothing or it may ring a bell and cause the trinket to vanish in a puff of smoke. Now, the magic here is that the shop guarantees that some combinations of button presses by you and your long-distance friend will cause bells to ring; other combinations will not.

It is impossible to find in physics a solution to how that information could be imbedded in the trinket itself. There must be some information flow (impossible because of the distance) or some very spooky properties of the interior of the trinket. These properties are exactly what quantum mechanics predicts. But, whereas we have an explanation for the pairing up of the goblets (prepacked and labelled), there is absolutely no explanation in classical terms for the properties of the trinket.

This seemingly instantaneous action at a distance was first raised as a paradox by Albert Einstein, Boris Podolsky and Nathan Rosen in mid-1930s. A photon has associated with it a *spin*, which may have two values, up or down (almost as if it could turn in either direction). The essence of the paradox is this: when two photons are created from (say) the collision of a particle and its anti-particle equivalent, their spins are created equal and opposite. But which has which spin is not decided. They can travel to the opposite ends of the universe and either photon can assume either spin. The crunch is that when one photon's spin is determined the other, at the other end of the universe, is also determined: no communication takes place (since photons travel at the speed of light this would be rather difficult anyway). Somewhat later, John Bell showed that the results could not be the

35

result of some hidden mechanism or concealed information buried within the trinkets.

This 'action at a distance' that quantum mechanics gives us is philosophically very confusing. Although these effects are instantaneous across the universe and seem to defy relativity, which limits the speed of travel to that of light, they cannot be used to transmit information. The universe is somehow entangled in past histories of its constituent matter. We have yet to find any ways in which this manifests itself at a macroscopic level.

So we find ourselves in an interesting situation. At the beginning of the century the simple classical picture of the world gave way to quantum theory. The Heisenberg Uncertainty Principle destroyed, forever, the idea that we might have perfect knowledge of any physical system. The non-local properties gave very unintuitive predictions about observations made at a distance.

At the close of the century we have seen the chaos revolution and the realisation that the classical physics of Newton had nothing like the strict determinism we always imagined it had.

CHAPTER 3

MILITARY AND DEMOCRATIC ROUTES TO COMPLEXITY

When we watch a military parade, we see squads of soldiers marching in perfect step. Their walking is *synchronised*. We don't need to track every soldier individually: we need only to know where the squad is and what direction and speed it's going in. With just this small amount of information, we can determine the movement of every soldier.

Soldiers are trained to march. But many natural systems show synchronisation like this without any external influence. On the other hand, animals such as ants cooperate, producing complex behaviour for the colony, but do not have this military precision.

MARCHING IN STEP

The laser is a modern embodiment of the ray gun of science fiction. But despite Ronald Reagan's Star Wars efforts, the laser's main uses are in technology, not war. The laser is now a common household device: all CD players, for example, use lasers to read compact discs.

Lasers have come a long way from science fiction. We have also come a long way from the theoretical turbulence the laser created. According to statistical thermodynamics, the atomic or molecular particles of a gas have a wide range of energies. This distribution tails off at higher values and is referred to as the Maxwell distribution. This distribution characterises *thermal equilibrium* brought about by collisions amongst gas molecules. Temperature is defined as the average molecular kinetic energy. In a laser, just before firing, the gas is far from such an equilibrium. The advent of the laser thus introduced a new system, not in equilibrium and with a vast distortion of the energy distribution.

The activity of a laser occurs in two stages: in the first (pumping) stage, light is applied to the laser gas, injecting energy; in the second stage the energy is released in a coherent pulse. In the pumping phase (we shall gloss over the technical details) gas molecules are raised to an excited energy state as they absorb light. In traditional theory they would shed this energy by collisions, the energy distribution would even out according to the Maxwell distribution, and the overall temperature would rise. But for a variety of reasons, amongst which is the rate at which energy is being pumped and trapped in the system, this does not happen and we end up with a population inversion where there are far more molecules in higher energy states than in lower—a lot more. The right trigger can now release all of this excitation energy in a single coherent pulse, the laser beam.

One of the prominent early theorists of the laser, Hermann Haken, is also one of the great unsung heroes of complexity. Haken's analysis led him to formulate the 'slaving principle', which he then proceeded to apply to many other systems, developing a field he called *Synergetics*, the science of *self-organising* systems. In fact Haken has hosted a series of conferences on Synergetics going back nearly two decades! The mathematics of Synergetics is quite difficult but we can explain the two core ideas. A complex dynamical system has many variables, but under the right circumstances some variables will enslave the others, just as the drill sergeant keeps the soldiers marching together. The system acts in unison under just these dominant variables. But for this to happen, noise—or in Haken's terms, fluctuation—is essential to give the enslaved modes enough plasticity to get into step. There are many simple systems that exhibit slaving, without us needing to look at systems that are obviously diverse in number and type of agents, such as ecologies.

Something as simple as a boiling liquid exhibits a synergetic phenomenon—Benard cell convection. The liquid self-organises into columns of hexagonal cells: in some columns liquid travels up from the bottom of the vessel, while in adjacent columns liquid travels down. Just think how extraordinary this is: there are no inherent boundaries in the liquid and these columns have formed spontaneously. Furthermore, they have a clear geometrical shape, again something in no sense obvious from the initial setup.

Marching in step is an example of rhythmical, coupled movement. Periodic motion, from pendulums to planets, is something we learn about in early childhood. Before the advent of quartz watches it was the fundamental component of a timepiece. Quartz devices, of course, are still oscillators, but the oscillations are hidden down at an atomic scale. A simple oscillation has just three properties: the frequency with which it returns to the same point in the cycle; the amplitude, which is a measure of the power in the oscillation; and the phase, the point at which the cycles begin.

We see interaction of oscillators in many places in everyday life. In an early differentiation of the sexes, Mary has her two small children Jack and Jill on the swings in the park. Jill rocks gently back and forth, but Mary gives Jack's swing just a little nudge at the high point and the swing gets bigger and bigger. But Jill does not want to be left out. Each time Jack passes she tries to grab him. As she touches him she slows him down and he speeds her up. Before long, both are swinging together. Later on we shall show how this situation of the swings together is an *attractor* in the joint dynamics of two systems. Part of the idea of an attractor is that it is easy to get in and hard to get out. Hence, often, only small amounts of energy are needed to bring about synchronisation of behaviour.

Several simple oscillators that can couple with each other in some way display an interesting effect: even if the frequencies are not all exactly the same and even if their phases are random, eventually they will lock together, with the same frequency and phase of oscillation. This remarkable effect was first observed by Christian Huygens, the inventor of the wave theory of light, in the synchronisation of clocks on a wall. They might start off all reading different times, some running fast, others slow, yet after a while they will all end up going at exactly the same rate. Adjust them to all read the same time and they will stay that way, although they may all be running at the (same) wrong rate. For this *phase-locking* to occur the clocks must be able to link up in some way: by putting them against a wall, their mechanisms transmit vibrations into the walls which serve to couple them together.

Phase-locking is a common problem in bodily activity. For instance, a common childhood game is to try to pat your head and rub your stomach at the same time: it's surprisingly difficult. One of the most difficult exercises for beginning pianists is to play different rhythms with each hand—part of the complexity underlying Chopin's superb miniatures. Scott Kelso, in his wide-ranging book, *Dynamic Patterns*, discusses many variations on this theme, including a test used by the Juilliard School of Music in the United States. The test is really to see how well a potential student can maintain a rhythm in the presence of interference from a metro-

40

nome. How well can the student resist the powerful drive to synchronisation? In fact music is full of synchronisations, from dancing to heavy beats to the difficulty of playing the polyrhythms of Stravinsky's *Rite of Spring*.

An even more subtle phase-locking happens among young women. Sue, Jill and Sarah share a house: they moved in at different times, had never met before, and each spends as much time with her own circle of friends as the three of them do together. Yet after a few months of living together, their menstrual cycles phase-lock. What is the coupling mechanism? One suggestion is pheromones, the sex hormones well known elsewhere in the animal kingdom. So the coupling is very slight, almost subliminal, but achieves synchronisation all the same.

As the sun goes down, as the town lights up and the revelry begins, a party of a different kind is just beginning. Away out in the wilds a firefly lights up, then another and another, until the sky is spotted with bright little dots of light. Then slowly these little pulsating dots start to phase-lock. Before long there is just one coherent flash as the fireflies turn on and off together. The coupling method is visual. So why do the fireflies synchronise like this? Could there be an advantage?—it might make it harder for a predator to focus on any one firefly. There might be an explanation of this kind, but in fact it isn't necessary. These dynamical effects occur spontaneously in oscillator systems and in fact are difficult to avoid, as we shall see now.

Animal locomotion

If you watch a cat moving around you'll see several different styles: at the slow end there is paw by paw stalking, a very slow walk. Beyond simple prowling there is a half-speed trot which finally becomes a gallop in

41

which front paws and back paws move together. Until you read a book such as Kelso's, it may never occur to you that a cat's movement might be anything other than under conscious control. Surely the cat decides whether it wants to trot or gallop. Not necessarily so, according to Kelso.

The transition between gaits can be accounted for by simple dynamical models. For instance, Kelso describes a simple experiment, finger walking. Try 'walking' on the desk with two fingers of one hand; that is, allowing index and middle finger to alternatively hit the desk. Now do both hands together, but slowly in antiphase, so that opposite fingers are down at the same time on the two hands. Easy? Now gradually speed up. At some point most people will suddenly break into the symmetrical pattern, like a cat galloping, where the fingers hit the desk together.

From ants to orcas

Ants present a source of fascination missing from many other insects. Although the destructive force of giant soldier ants is awesome, it's the complexity of their society that really grips us. How do creatures with such small brains and minimal senses achieve complex domestic arrangements and coordinated foraging? In some ways our education tends to mask the simplicity of what goes on.

An experiment with young children illustrates this. They were asked to work out how to search for gold on an alien planet, having got there by spaceship. All their solutions emphasised communication with the spaceship and central planning and direction. But ant colonies don't work like this. There is no central planning. But ants are very cooperative via signals (scents) between individuals.

Edmund Wilson is an expert on ant and ant colony behaviour. However, in 1975 he risked widespread opprobium by publishing a book entitled *Sociobiology*, in which he attempted to extend what he knew of ants and other animals to the anthropological domain. The parallels in behaviour across widely different species are amazing. In some of the most exciting and daring wildlife photography in the series *Life on Earth*, an orca, a killer whale, is seen taking a seal off the beach and then proceeding to play with it in just the way a domestic cat plays with a mouse. Same behaviour pattern, but the only thing in common between the two predatory animals is black and white markings! This is in some ways the essence of a science of complexity. What surface phenomena can we strip away to get down to the underlying forces?

Craig Reynolds' 'boids' are the simplest artificial bird imaginable. They are just simple icons on a computer screen with a few simple rules of interaction: they shouldn't get too close, should fly at roughly the same speed, and so on. Yet a flock of such boids flies in formation just like a flock of geese, the leader changes at regular intervals and they fly around obstacles just like the real flock. So here we have a phenomenon which looks the same either in a very simple simulation or in a complex organism. Is this then a scientific model? Is this the way to model biology in just the way that physics starts off with simple models? Such philosophical considerations will concern us later.

To see this philosophical issue more clearly, pseudopotential theory in solid-state physics illustrates the transition from simple to elaborated model. The first models of the properties of metals were derived from very simple quantum-mechanical models. A metal contains many atoms in a fairly regular lattice; the atomic nuclei remain anchored to the lattice points, as do most of the electrons. But some of the electrons are essentially free to roam around. Unfortunately, quantum mechanics doesn't allow any simple cancellation of electrons (negatively charged) with protons in the nucleus (positively charged). Nevertheless, the early models of solid-state physics threw away these unpleasant considerations and simply ignored almost every salient interaction in a metal! The results were excellent, but nagging doubt remained as to why this should be so successful. Deeper understanding acquired over subsequent years led to a new concept, the *pseudopotential*, which was not only quantum-mechanically rigorous but also explained in a simple way why the earlier models worked so well.

Returning to boids, we have a simple, successful model of very complex organisms. But theoretically we hit a snag immediately. In physics, measurements are frequently extremely precise, facilitating detailed comparison of theories. Biological data rarely approach this precision, making the assessment of models much more difficult.

The boid model is an algorithm, a computational procedure for determining outputs (movement) from inputs (proximity of other birds). The bird's brain achieves this same approximate mapping between these two domains. But as we discuss later, it is hard to say that the brain implements a particular computer algorithm. How we go from these simple models to an explanation in biology is still far from clear.

Are animals conscious? We all tend to assume that our pets have identities and some sort of consciousness. In fact Donald Griffin, Harvard ethologist, details extensive human-like behaviour in many animals. Beavers, for example, build dams in teams but display adaptable behaviour in emergencies. Griffin cites examples where they will *reverse* their normal practices where necessary in situations they have never seen before.

The beaver of course has a much bigger brain than any insect and some level of sophisticated communication. But now we can turn the question the other way around. How much of beaver behaviour can be explained by simple mechanisms as opposed to conscious decision making?

A current craze, which originated in Japan, is the *Tamagochi*. This is an *extremely simple* processor chip inside a little egg-like case with a computer screen. It's a computer pet. It whinges when it wants 'feeding' or needs 'attention' and has a handful of other pet-like behaviours. You need only a few attributes of cuteness, it seems, to make a machine seem to come alive.

History seems to have a cyclical character. Peace and war alternate, customs come and go, political structures wax and wane, with no still point in the turning world. Predicting the course of human society (or, even better, changing it) has so far been the stuff of novels, such as the theme of Isaac Asimov's massive *Foundation Trilogy*. In Asimov's novels the Selsdon Plan was a detailed map of social activity within the galaxy, of mathematical precision provided the numbers were large enough. Of course it didn't work because of evolutionary events unpredictable at the outset. It's not obvious that we are any further advanced.

It is disturbing, however, just how little use analysis is in controlling human events on a large scale. The 1990s war in Yugoslavia, in which the country blew itself apart, is perhaps a sad case in point. Here we have a fully developed Western society, as technologically advanced as anywhere. We have universities, television, free access to books and newspapers. Above all we have intense international scrutiny, discussion and support. All of this has achieved nothing.

Thus we might speculate that the way forward is not to try a predictive model at the macroscale level or to focus on outcomes—such as the strife in Yugoslavia. Instead we have to focus on the local interactions: change these and the rest will follow. We might say that there is pressure towards large-scale behaviour—war, peace, anarchy—which is driven from within and very difficult to control from without.

THE MORE THE MERRIER

We have seen that interactions produce complex behaviour. It doesn't really matter what the interaction is between. It could be something as complex as a person, as simple as an ant, or not a living thing at all. It is convenient just to call it an *agent*. Now we should look in more detail at how all this comes about. So for the time being we are going to leave animals and human beings and retreat into abstraction, mathematics without formulae.

We naturally tend to think of individual human behaviour and of human society as the most complex on earth. The human brain, as John Eccles, Nobel Laureate in Medicine, remarked, is the most complex thing in the universe! Yet one of the salutary lessons of the study of complexity is that it can arise from seemingly very simple beginnings. Perhaps we need a definition of complexity, but this will have to wait for a while, since a good definition is surprisingly hard to create. We need some experience of complex systems first.

Let's begin with Grid World, an imaginary civilisation on Jupiter, where the gravity is so strong that the inhabitants, Griddies, are as flat as pancakes and are literally unable to move. In fact they are rather like plants, but gravity stunts their growth. Fortunately, in Grid World the nutrients come from underground through a series of pipes spaced out at regular intervals in a nice orderly grid. Unfortunately, food is limited, and Griddies

46

quickly die if they are surrounded by other Griddies on all sides.

The mating ritual in Grid World is pretty strange. It's a ménage à trois, but the act of reproduction is three-way with the three parents lined up in a triangle, and the offspring immediately appears in the middle, ready to tap into a food source. No suckling by the parents here. Griddies are also quite social; a Griddie left by itself soon dies. Griddipologists, the scholars of Griddie society, will tell you of great cultural diversity. Just as starships restrained by the speed of light go through generations of astronauts before reaching destinations light years away, so individual Griddies don't witness the great distances travelled by their offspring. But certain patterns of Griddies move like tribes from one place to another, preserving exactly the same shape as they move.

We won't go into any more detail about their society and customs for the moment, but instead leap forward to the seminal work of Griddie scholar Johann Konweg. The breakthrough was the realisation that, despite the depth and sophistication of the Griddie soul, the patterns of activity in Griddie society could all be modelled, on a piece of graph paper with a pencil and an eraser: a Griddie was just a square coloured in. Of course Griddie scholars use computers to look at social patterns these days, and this is exactly how these figures were drawn. We shall come back to Grid World later, because this extremely simple system displays some of the most profound ideas of complexity.

John Conway's Game of Life, a purely mathematical creation, is the underlying model for Grid World. But the message is a profound one. Complex behaviour in social systems might have very simple underlying structure. The drive to study complexity comes from the hope that there are general principles underlying many or all complex systems.

The Game of Life is very much a visual experience. Computer graphics have been the key to watching complexity unfold and we turn next to the use of computer simulation and models.

OF WORLD MODELS AND MODEL WORLDS

In 1968 a small group of scientists met in Rome to discuss the future of humankind. This was no idle talkfest. The aim was to determine just how serious were threats such as overpopulation and pollution that now hang over our future. Out of this meeting there arose an informal organisation—the Club of Rome. In 1970 the club commissioned scientists at the Massachusetts Institute of Technology to develop a simulation model of the limits and consequences of continued worldwide growth. The model incorporated modules dealing with population, agriculture, resources, industry and pollution.

The result were summarised in the 1972 book *The Limits to Growth*. The outcome was startling. Their 'standard run', which made forward projections from current trends, predicted a world catastrophe by the year 2050. Food and resources would be in short supply, pollution would peak and population would crash.

Needless to say, such dire predictions caused a sensation. Numerous critiques appeared, pointing out flaws in the model. Yet for all its faults the Limits to Growth model achieved its greatest aim. The stated goal of the modellers was:

> [To] foster understanding of the varied but interdependent components—economic, political, natural and social—that make up the global system in which we all live; to bring that new understanding to the attention of policy-makers

and the public worldwide; and in this way to promote new policy initiatives and action.[8]

In this sense the project succeeded. Up until that time warnings of a global environmental crisis were generally treated as fantasy. The Limits to Growth model started people thinking seriously about the future of the planet as a whole.

The Limits to Growth exercise is a good example of using a simulation model to study the behaviour of a complex system. Simulations are computer programs that recreate features of the real world. For instance, as we discuss elsewhere, the mathematics of astronomical bodies is unsolvable for as few as three bodies, such as the Earth, Moon and Sun. So the only way to model the formation of (say) a galaxy is to represent it in a simulation.

The importance of simulation models is that they allow us to perform *virtual* experiments. This ability is especially important where it is just not possible to perform the experiment concerned for real. For instance, we cannot heat up the Earth by several degrees and watch what happens. Even if we could, the risk of destroying civilisation in the process would make it a risky experiment, to say the least! However, by running a simulation *scenario* we can look at the consequences of runaway global warming without the risk of a real experiment. Aircraft manufacturers now prefer to use simulations of aircraft design before placing a design inside a wind tunnel. Likewise, trainee pilots can safely practice takeoffs and landings in a simulator before they have to face the real thing.

Previously we saw how John Conway's computer game LIFE might be adapted to model the hypothetical Griddies on Jupiter. These models are examples of *cellular automata*, or CAs for short. They are one solution

49

to the problem of simulating processes that are spread out across space. A CA is really a grid of cells that are spread out in 1, 2 or 3 dimensions.

In two dimensions we can represent a landscape by carving up space like a chequerboard. Each cell represents an area of space. The cells are all computers (*automata*). They are identically programmed. So at any given location the state of the corresponding cell tells us something about the objects or properties to be found there.

For example, the CA grid might represent an area of the landscape and its cells contain information about such features as the local plants, elevation and soil moisture. If we program the cells appropriately then the model might simulate processes such as tree growth and seed dispersal within a forest.

Perhaps the most important feature of a CA is that the cells in the grid interact with one another. For example, if a plant dies then to replace it we can simulate seed dispersal from all the plants lying within some *neighbourhood* of the cell concerned. These neighbourhood interactions make CA models non-linear.

Probably the best known CA models are simulations of global climate change. The so-called *general circulation models* (GCM) represent the world (or particular regions) as a large grid. Each cell in the grid represents an area on the earth's surface and the atmosphere above it. To model global climate they simulate major processes such as air movement and heat and moisture transfer within and between cells. The cells are classified according to the dominant type of environment present within them (e.g. ocean, desert, pastoral, temperate forest). Critical parameters, such as albedo (the fraction of solar radiation that is reflected) and evaporation, are assigned values for each type of environment.

In the Adams novel, *The Hitchhiker's Guide to the Galaxy*, the Earth itself is portrayed as the ultimate

50

computer. Unfortunately real-life computers have somewhat less capacity! Even on a large, modern
supercomputer, model builders have to make many
simplifying assumptions just to reduce the problem to
one that the machine can handle. It is just not possible
to represent the entire system accurately. To understand
what this might mean in a global change model, try to
imagine a giant leaf, 100 kilometres by 100 kilometres
in area. This is literally how some simulations represent
the forests within a cell of their grid. It will come as no
surprise then to learn that the models can be extremely
sensitive to these sorts of assumptions. This is one of
the reasons why different studies of global change can
give inconsistent results.

Given the sort of unreliability we have just alluded
to, many people might throw up their hands and say
we should give up (many do!). But all is not lost. As
with the Limits to Growth model, simulations are still
useful even if they do not accurately predict what will
happen in a system.

To understand this point let's take the problem of
predicting the spread of a wildfire. On the face of it this
looks to be a much simpler problem. Here we set up a
grid in which the cells represent fuel. In different
landscapes this fuel may be grass, or bushes or trees.
To simulate a fire we set one or more cells 'burning'
and the computer works out how the fire spreads from
one cell to another.

So far so good. Unfortunately, fire-spread models
come unstuck when we try to predict what happens in
a real fire. This is not because we don't understand
fire-spread at all. It is because of two problems: inadequate data, and the sensitivity of the system. Suppose
that we want to model the spread of a fire through grass.
Then the essential problem is that we need an accurate
map of where the grass is. Now, a grassland may look

51

Figure 3.1 Fire control scenarios in a CA model of the spread of a wildfire. The map is a digitised aerial photograph of a landscape, with roads, trees and fields. The fire-front is seen here as a black line.

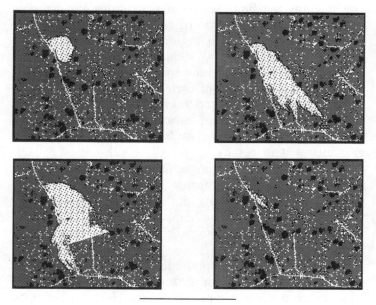

like a flat carpet from a distance, but it is composed of millions of individual grass tussocks of varying size, density and composition. Even to gather this sort of data for an experimental area is a huge job, and unfortunately wildfires never oblige by burning on experimental lots! But even if they did there is the problem of fuel moisture. The danger from a fire depends not only on how much fuel is available but also on how dry it is. Even if we were to go out and measure the moisture, we'd find it changing constantly. Also changing is the wind speed and direction, which often drives a fire-front.

Compounding this problem of poor data is the problem of sensitivity. The pattern of fire-spread is simply very sensitive to fuel load and moisture. If the grass is slightly denser and drier in one area then the fire will spread faster. The effect is even worse if the fuel is patchy, with (say) large gaps between bushes or trees.

We have described this fuel problem in some detail to make it clear that in many cases the exact pattern of fire-spread is literally unpredictable. However, this does not mean that models of fire-spread are useless. Far from it. They are simply not good for on-ground fire-fighting. Their real use is in asking what-if questions. What if there is a sudden wind change? Would a fire-break along a ridge be adequate? Would off-season burning reduce the risk for campers in the summer?

From virtual worlds to artificial life

This serious-minded 'game-playing' with cellular automata and other systems is one aspect of a booming new area of research called *artificial life*. Artificial life (or *alife* for short) aims to understand living systems better by abstracting them and simulating them in computers and other media. Besides cellular automata, alife looks at many other kinds of systems as well. Many of the ideas we shall encounter in this book really fall under the heading of alife. For instance, we can enrich a CA model by laying over it roving *agents*. These are effectively programmed automata, but instead of remaining in one spot they can move around, just as (say) a deer might wander through a forest, or a fish swim around in a coral reef.

An important method in science is to generalise from specific cases. Traditionally this has been the role of mathematics. For instance, if we can write an equation

53

Figure 3.2 Examples of 1-dimensional cellular automata. Each system starts off as a row of cells (at the top of each block), with a random arrangement of black and white cells. The rows below show the new states of the cells in the CA. That is, each column follows the history of changes in the colour of a single cell. The example here represent the classes identified by Wolfram. At top left is a CA in which a brief series of transient states leads to a constant system (all cells white). At top right is an example of cyclic behaviour. The examples on the bottom row show more complex, chaotic behaviour.

to model the growth of a population, then by studying the properties of that equation we learn not just about the population in question but also about any other system for which that equation is appropriate.

The same holds true for simulation. As we have seen, cellular automata provide models for many different systems. So from games like LIFE we can learn a lot about the behaviour of many different kinds of system.

Research on CAs really has several aims. On the one hand the models can be seen as artificial worlds. But at the same time they can be seen as models of parallel computing. As we have noted, each of the cells in a CA is a little computer. So regardless of what natural system we simulate with a CA, its grid of cells also mimics a parallel computer. It is recognition of this fact that motivates a lot of the research on CAs. A couple of decades of research on cellular automata have taught us a lot about systems of interacting cells. Stephen Wolfram, for instance, found that the behaviour of CAs fell into several general categories. All of these are visible in the computer game LIFE.

The first, and simplest, behaviour is for a system to 'freeze' into a fixed state, after which it never changes. In the game LIFE, for instance, several patterns of living cells rapidly die out, leaving behind an unchanging sea of dead cells. Others form fixed patterns that never change.

Slightly more interesting are patterns that fall into cyclic behaviour, periodically changing between several different patterns. In LIFE, a row of three live cells blinks: first it is vertical, then horizontal, then vertical again, and so on.

More complex is chaotic, aperiodic behaviour in which the system never settles down to a constant pattern. Finally there are complicated localised structures, some of which may propagate.

We can see examples of all these types of behaviour in a simple one-dimensional string of cells as it evolves through time. Because it is just a line we can lay each successive pattern that we see in the system side by side vertically, to form a two-dimensional pattern. In these pictures the vertical axis represents time. Any column in the picture shows us the history of changes in one particular cell. This means that a column of constant colour represents a cell that falls into a fixed state. A column of alternating colours indicates a cell that cycles. The other two types of behaviour are seen as larger scale patterns that *emerge* from the behaviour of the cells.

Subsequent work by researchers such as Chris Langton has shown that the different classes of behaviour tend to be associated with how complex, how 'hot', the CA model is. At first this idea of temperature for a model may seem obscure and technical. In fact it's fairly simple.

As we saw earlier, the behaviour of cells in a CA depends largely on its neighbours. One way to describe temperature is by asking whether the neighbourhood of a cell forces it to change state. If most arrangements of cell states in a neighbourhood will leave a cell unchanged then the model is very 'cool'. Cool models are usually associated with the first two of the above types of behaviour. When the model is run, the few neighbourhoods that force a change of state do so, after which the pattern of cells freezes. In a very 'hot' model, most neighbourhoods force cells to change state, so the pattern in the CA bubbles away furiously and can never settle down. So hot models lead to the third type of behaviour. The final type of behaviour seems to be associated with models that lie in between the others, in the border region between very ordered behaviour (fixed states and cycles) and chaotic behaviour.

This final type of model seems to exhibit the most interesting behaviour. That these models seem to lie in

Figure 3.3 **Phase changes in the behaviour of a CA model. The three models shown start off with the same random arrangement of black and white cells but soon produce the patterns shown here. The model on the left has low complexity and is 'cool': it changes the colour of only a few cells, so the system 'freezes' before any interesting patterns emerge in the array. In the model on the right, its behavioural complexity is high and it behaves much like a 'hot' gas: it changes the colour of almost every cell. Its behaviour is essentially chaotic. No patterns can emerge because the grid is constantly 'boiling'. The central, medium model lies at the 'edge of chaos': its behaviour is richer than the model on the left, but much 'cooler' than the model on the right, so pattern elements, such as the lines visible here, do emerge in this model.**

COOL   MEDIUM   HOT

a region between order and chaos has led to the provocative name *edge of chaos* to describe them.

This idea will occur through the remainder of the book. There are special properties of *connectivity* and *dimensionality* characterising this model. In fact, complex adaptive systems may evolve so that their behaviour lies at the edge of chaos, a startling conjecture that is one of Stuart Kauffman's many brilliant insights and can be found, with many other similarly profound theories, in his book *Origins of Order*. Kauffman is well known in

57

alife circles. A distinguished biologist, he has moved in recent years to the study of complexity, principally at the Santa Fe Institute.

Chapter 4

The Imbalance
of Nature

On 5 November 1983 the *Weekend Australian* carried a startling headline: REEF PARK A FARCE IN FACE OF STARFISH THREAT.

The 'Crown' of Thorns' starfish—*Acanthaster planci*—was destroying the Great Barrier Reef. The largest coral reef in the world, it seemed, was doomed.

Over the next two years the media carried many similar stories about the killer starfish. TV news showed footage of distraught tour operators desperately removing starfish by the boatload. Outraged scientists appeared on screen deploring the lack of government action. To the person in the street it seemed that we were witnessing the demise of one of the great natural wonders of the world. Millions of killer starfish were swarming over the reef eating the living corals. When the corals were gone, the fish and other animals departed. What was left was like an underwater desert, grey and lifeless.

Within the scientific community a savage controversy erupted over the nature and cause of the outbreak. One side claimed that we were witnessing the end of the Great Barrier Reef; the other side insisted that the outbreak was simply a natural phenomenon. For years the scientific debate was as fierce as the starfish themselves.

WHY IS A STARFISH LIKE AN ATOMIC BOMB?

As is usual in such polarised debates, the truth, as it eventually turned out, was not nearly so simple. With an election looming at the end of 1984 the Australian Government threw money at the problem. It commissioned AIMS, Australia's prestigious Institute for Marine Science, to carry out a survey of the problem. Roger Bradbury, the Institute's chief ecologist, led a team of more than 50 young scientists who scoured the Great Barrier Reef (GBR) looking for starfish. After several years of hard work the answers began to emerge.

The GBR, it turned out, was not dying after all—not yet, anyway. Even so, an event of enormous magnitude certainly was taking place. To understand this event we must first understand the nature of the GBR.

Most people think of the Great Barrier Reef as a gigantic wall of coral that runs all the way down the Queensland coast—nature's equivalent of the Great Wall of China. It is the obstacle that nearly wrecked Captain Cook's ship, *Endeavour*, during his voyage of discovery in 1770. Over 2000 kilometres long and millions of years old, it is the largest coral reef in the world. It is also the home for thousands of marine species.

The GBR is all of the above things and more. But one thing it definitely is not is a solid wall of coral. In reality it is a chain of reefs, thousands of them, surrounding islands, atolls or shallows that litter the coastal waters. The GBR is a series of reefs, each one separated from the others by deep water. Up until the time of the AIMS study everyone studying the problem had been mesmerised by the scale of destruction on individual reefs. Bradbury realised that, to understand what was happening, his team had to study COT (Crown of Thorns) outbreaks as a system-wide phenomenon.

A coral reef is a complex ecological community. It consists of hundreds of species, including fish, sponges, crabs and many other kinds of plants and animals. But ultimately all of them rely on the coral reef for their existence. The corals themselves are communities made up of millions of tiny organisms. A coral is not a single organism at all but a symbiotic relationship between a plant and an animal. The massive coral domes and branches we see are the by-products of this symbiotic relationship. All of their remarkable activity depends on sunlight and on water. The algae need light to function, so corals cannot grow in water that is more than 30 metres deep. Nor can corals survive prolonged exposure to the air. So they are confined to the shallow waters near the coast.

For bottom-dwelling corals, the deep water between islands is akin to a desert. No corals can grow in the darker waters, so no reefs form. Where coral reefs do abut deep water it is because their reef-building has kept pace with rising sea levels over thousands of years.

So when TV cameras show pictures of total destruction on the reef, what they are showing is the destruction of a single reef: a local disaster. Reefs a few kilometres away may be totally unaffected. At the time that the COT problem first came to notice, no one knew just how many reefs were affected, or how badly. Most important of all, no one knew whether the reefs would ever recover. Alarmists claimed that the reefs were doomed, but was that really the case?

The AIMS survey soon revealed some intriguing patterns. First, the destruction was patchy. Some reefs were devastated, but others were untouched. Moreover, there were no half measures: either a reef was totally destroyed or there were virtually no starfish at all. Nowhere were there large numbers of starfish living in harmony with the reef. This was one case, it seemed, where the balance of nature did not exist.

What was going on? Did some reefs have special properties that made them immune? Were some reefs more susceptible to attack than others? More surveys, and accumulating historical data, began to reveal the answer. When the scientists plotted maps of outbreaks against latitude they made a startling discovery. It appeared that a wave of starfish outbreaks was moving down the GBR from north to south. This pattern soon came to be known as the 'Reichelt Wave', after Russ Reichelt, a member of Bradbury's team and later director of AIMS, who first noticed the pattern. The wave of outbreaks moved south over a period of years. The wave pattern was weak in the north, where outbreaks seemed to be more sporadic, but the pattern became progressively stronger as it moved south.

Once scientists had found a systematic pattern in the outbreaks, they began to appear more and more like a natural event. As the scientists gathered further historical data, it became apparent that the GBR had been subjected to waves of outbreaks many times in the past. It even began to appear that they might occur with a predictable frequency. Each wave of outbreaks would run its course in a few years, after which the GBR would gradually return to its usual healthy state.

Having identified the pattern, the scientists next turned their attention to the process. Several crucial questions still remained unanswered. In particular, how did the outbreaks spread? Did armies of starfish march across the ocean floor from one reef to the next? Tracking individual starfish is notoriously difficult. Unlike fish or birds, you cannot tag a starfish. Try it, and the starfish simply ejects the metal from its body.

Despite these difficulties, the biologists soon had their answer. During the summer spawning period the starfish released their young. The waters filled with clouds of minute larvae. They spread in such vast quantities that even the hungriest predators could eat their fill and yet have no noticeable effect on the numbers. Within hours the clouds drifted away from the reefs where they formed and moved out into deeper waters. Currents then picked up the larvae, driving them away from their source. Oceanographers have long known that currents wash the entire length of the GBR from north to south. These currents now carried the larvae south. Within days millions of them had settled. Another reef was doomed. The cause of the Reichelt Wave was now clear: outbreaks on any given reef eventually lead to a larval cloud, which then seeds outbreaks on other reefs, and so on.

The only remaining question was how the outbreaks started in the first place. Obviously the process has to be seeded by outbreaks in the waters north of Cape York, but how? Without massive influxes of starfish larvae, why should an outbreak suddenly start? Is it chance? Or is it something more serious? The most contentious theory is that overfishing reduces the predators that normally feed on starfish larvae and young starfish. The issue is hotly debated to this day.

At least now we know how the outbreaks spread. At the time of writing, fresh outbreaks are again being reported in northern reefs. If the above theory is correct, then scientists should be able to observe another wave of outbreaks over the next few years. Most importantly, the process is now predictable. We know what to expect. We need not be quite so alarmed the next time starfish begin appearing in a prime tourist spot. Perhaps we can even prevent it.

As it now appears, the COT outbreaks are what scientists refer to as an 'epidemic process'. When one reef succumbs it is akin to being infected with a disease.

Moreover, it infects other reefs in the same way that colds and flu spread through a schoolground or an office.

All through history, humans have been subjected to plagues of one kind or another. The AIDS epidemic is but the latest. During the Middle Ages, Europe was ravaged time and again by plagues that swept across the continent, usually starting from somewhere in the Middle East. Towns were worst affected. All it needed was for a single infected individual to arrive in town. By the time the unsuspecting inhabitants knew what was happening it was too late. Within a few weeks trundle carts would be plying the streets collecting the dead by the hundreds. Many rural villages, knowing their peril, would close themselves off from the world and take the extreme measure of killing any strangers unlucky enough to arrive seeking shelter for the night.

So the COT outbreaks are a typical epidemic process. Many other processes act in this way too. For instance, we tend to think of fire burning through a landscape as almost a living, growing thing. Yet fire too is an epidemic process. Fire spreads from one tree, one bush or tussock of grass to the next. If the bushes are too far apart, then the fire cannot spread. Though less obviously, water seeping through soil, and weeds spreading from paddock to paddock, are also epidemics of a kind.

But perhaps the best known, and indeed most infamous, epidemic is the nuclear chain reaction that sets off an atomic bomb. An atomic explosion occurs when a neutron collides with a uranium or plutonium atom, causing the atom to split in two. In the process the atom fires out two new neutrons. Each of these goes on to hit other atoms, which also split and send out still more neutrons. The explosion is caused by the resulting heat, which is released at a rapidly increasing rate as the number of splitting atoms doubles, quadruples and so on. The process spreads in essentially the same fashion as any epidemic, including the spread of COT starfish.

And that is why a starfish, or at least an outbreak of them, is like an atomic bomb. So what?, you may ask. How does it help us to know that a starfish is like an atomic explosion? Surely the two things are even more different from one another than chalk and cheese?

In truth it does matter that starfish outbreaks, plagues, fires and atomic fission are all related processes. For in understanding this similarity we can learn more about each one than we ever could by studying it on its own. For instance, before they could build an atomic bomb the nuclear physicists at work on the Manhattan Project first had to learn about the properties of chain reactions. Some of these discoveries apply directly to other epidemics. For instance, to work at all an atomic bomb has to have a certain 'critical mass' of fuel. If the lump of uranium is too small, then nothing happens. The alpha particles emitted from a dividing nucleus simply shoot off into empty space, missing all the other atoms completely. The bomb makers exploited this property by dividing the nuclear fuel in two. Each piece by itself was too small to explode, but when slammed together by a conventional chemical explosion the two formed a critical mass.

For similar reasons the classic approach to infections is isolation. Isolate the cases and you avoid forming a 'critical mass' of human 'fuel'. The point is that, in identifying this common property, one system can offer insights about another. For instance, perhaps some critical number of COT outbreaks in northern waters is necessary before a wave of outbreaks can get under way. If so, then one way to avoid starfish outbreaks is to reduce the incidence of those initial outbreaks. Curtailing fishing may be one answer. Short of that, however, applying preventative containment to seed outbreaks is likely to be far more cost-effective than a Bandaid approach of trying to rescue dozens of high-profile reefs once the process is under way.

Phenomena, such as epidemics, that characterise complex systems permeate the whole of the natural world. It is possible that they even include mental phenomena. Are flashes of insight and bright ideas examples of neural epidemics? Such questions illustrate the extreme diversity of complex systems. Yet, given this diversity, there seem to be underlying patterns which beg for some global theory or paradigm. An almost universal characteristic of a complex system is a set of agents of some kind which interact in some way. These agents may themselves be complex, but the properties of the system are not easily predicted from a knowledge of the agents and the way they interact. Only step-by-step simulation of the system will reveal its behaviour.

One very important parameter of complex systems is connectivity: which agents talk to which others. What is interesting, and extremely powerful as a theoretical tool, is the discovery that it is just the existence or lack of connections that is significant, not their strength or dynamics. We look at the implications of this more closely in Chapter 6.

COMPLEXITY IN THE LIVING WORLD

One of the greatest challenges facing humanity today is to manage and conserve the world's living resources. Enlightened management has to be based on accurate, up-to-date information. The first problem is simply to learn what plants and animals are out there.

How many species are there? Science is not able to answer this simple question with any certainty. Estimates by different methods put the world's biodiversity at anywhere from two million up to 100 million species. A best guess is probably in the vicinity of ten million. The main reason why we are so vague about how many species there are is that only a tiny proportion of the world's plants and animals have ever been described scientifically. One UN estimate is that the task will take at least another 600 years to complete. The fear is that in rainforests and other ecosystems, whole communities of species are being wiped out before we even know they exist.

One way to estimate species numbers is to scale upwards from what we know. When scientists survey a new area for the first time, they are likely to find many new species. Later surveys inevitably find fewer and fewer new species as the stock of new species dwindles. This is a case of diminishing returns: it takes a lot of effort to find the last few species. On the other hand the rate at which new species are found falls off in a predictable way. So if we compare effort with the rate of discovery, then it is possible to predict how many species are still unknown. This kind of prediction indicates that there are relatively few unknown species in some well-studied parts of the world (e.g. most of Europe), but in other places there are still areas largely untouched. Even in developed countries, there are many blanks on maps of regions surveyed for biodiversity. In Australia, for instance, scientists have actually produced maps to show where specimens were obtained. For central Australia, these points formed a perfect map of the roads through the Australian outback! The rest of the continent is largely unsampled.

Many national and international projects are under way to compile lists of species for different parts of the world. One source of data is to catalogue specimens in taxonomic collections. This may not seem like an onerous task. However, to understand just how enormous a challenge we face in databasing biodiversity it is worth looking more closely at some examples, so we can know what is involved.

The Australian National Insect Collection is housed in a large building in Canberra. The main collection resides in row after row of cabinets that are spread over several floors. Each cabinet contains many drawers, and each drawer holds dozens of specimens. In all there are over eleven million specimens in the collection. Electronic cataloguing of this vast array of specimens began in the early 1990s with a single technician entering data off the specimen labels. It was estimated that the technician would be able to capture data for about 50 000 specimens in a single year. This effort sounds impressive until we consider the size of the collection. At that rate it would take over 200 years to record all the specimens!

Added to the sheer problem of recording the mountains of data already available is the problem of making sure that the data are reliable. Errors can creep in at any stage of the recording process. The original collector could make a mistake about where the specimen came from. The specimen could be misidentified. Likewise the scientific name could have changed and not been recorded. Then there are typographical errors, either in writing the original label or in entering it on a computer. The most useful data about location are the latitude and longitude. Early scientists were not aware of the needs of modern science. So crucial data such as location are sometimes missing. More often they are given in some off-the-cuff manner, such as 'five miles north of Gundagai'.

A little anecdote might help to drive home how serious all these errors can be. One of the first data sets to be compiled from specimen records, as described here, were lists of endangered species for Australia. When these data were first plotted as points on a map, it was found that several hundred specimens had been mislocated in the ocean! If this sounds bad, remember that these records were relatively easy to fix. Even more worrying than points mislocated in the sea are points that are wrong but not *obviously* out of place.

WHAT IS ECOLOGY?

Ecology is the branch of biology that looks at communities of plants and animals in the environment. Order in ecosystems arises from two different sources: the external environment and internal interactions. Ecology has different ways of looking at these two aspects of ecosystems. One way, *autecology*, looks at each species and how it interacts with the environment. What is its distribution? What sort of climate does it prefer? How does it reproduce and spread? Besides looking at interactions between organisms and their environment we also need to consider relationships between different kinds of organisms. *Synecology*, the study of living communities, looks at the ways in which populations interact.

The environment shapes natural communities in many ways. Plants cannot grow where the temperature is too extreme; animals cannot live where there is no water. There are endless variations on this theme. Because most environmental properties vary from one place to another, landscapes are a central concern in ecology. For instance, the temperature grows colder the higher one goes up a mountainside. Tropical heat at the foot of a mountain can give way to freezing cold at the top. This change gives rise to effects such as the alpine tree-line. In similar fashion, rainwater quickly runs off hilltops and into gullies where it becomes concentrated in creeks and rivers. This is why rainforest tends to be concentrated in valleys, whereas trees that prefer dry soils, such as pines and eucalypts, congregate on upper slopes and ridges.

Not only do properties of the landscape affect what can live where; processes that go on in the landscape have a big effect too. Floods, droughts, storms and fires all limit the kinds of plants and animals that can live in any given area. Fire in particular is an extremely powerful force for shaping communities in a landscape. In many parts of the world, native peoples deliberately light fires as a tool to manage the landscape for their own ends. One effect is to clear the underbrush, making hunting much easier. Another is to promote the fresh growth of grass and edible plants as food for game species.

The other major force that shapes ecosystems consists of the interactions between organisms. These interactions are often less obvious than environmental factors but in the long run have just as great an influence over what lives where. When we think of interactions between different living things the images that spring to mind tend to be lions chasing zebras, cattle eating grass or even viruses infecting human beings. As we shall see, these kinds of interactions certainly are important. But interactions between plants and animals of the same species are equally significant.

Cooperation and competition

Within any population, individual plants and animals interact in many different ways. The lyrebird dances to attract a mate. Male sea-lions fight each other for control of territory and females. Monkeys suckle and rear their offspring. Wolves cooperate to hunt in packs. Let's look briefly at some of these interactions more closely. We are all so familiar with the idea of animals raising their young that we tend to assume that parental care is a universal feature of the natural world. But this is not so. Most frogs, for example, lay their eggs and then forget them. Baby tadpoles are left to their own devices.

Parental care is really just one of many strategies that organisms can adopt towards reproduction. Ecologists have found that, in the struggle to survive in the environment, reproductive strategies lie between two extremes. One extreme strategy is to produce lots and lots of offspring, spread them far and wide, and hope that some will survive to adulthood. Locusts, starfish and other plague species adopt this strategy. So do many plant species. They rely on the odds that at least some offspring will survive. A very different strategy is to produce few offspring but to devote lots of time and effort to ensuring that they survive. Parental care—as we see in humans and other social animals—is one strategy of this kind. Other examples include cuckoos placing their offspring in the nests of other birds, and wasps laying their eggs in the bodies of living insects. Likewise fruit trees produce their seeds inside fleshy containers, which serve both to attract animals that disperse the seeds and to provide nutrients for the seeds when they germinate.

Another familiar kind of interaction between organisms is competition. Living things compete against one another for food, living space and mates. This competition provides the driving force for natural selection. More individuals are born than can be supported by the available resources. The competition for limited resources weeds out individuals that are poorly adapted and favours individuals that are *fittest* (best adapted). The crucial point is that the fittest individuals in a population tend to be the ones that survive to maturity and that then reproduce and pass on their genes to the next generation. So the best genes tend to be passed on from one generation to the next. This is natural selection.

Interactions between species also play a major role in shaping the complexity of nature. Just as competition occurs within a species, so too does competition between similar species. What are the effects of this competition?

One of the early ideas put forward about this was the *competitive exclusion hypothesis*. This theory proposed that whenever two populations compete with one another, one population will drive the other to local extinction. That is, one species disappears from the local area.

The above theory agrees with many field observations and correctly predicts what happens in laboratory experiments when (say) different microbes are left together in a flask. However, so many exceptions have been found that the idea has become largely discredited. Why have lions not driven hyenas to extinction? How are so many similar species of trees able to coexist? The answer seems to be that competition is not always as intense as it inevitably is in a closed laboratory flask. Also 'competing' populations are not always really competing. In the real world populations are spread across a landscape. One way in which similar species coexist is by dominating different territory. When this happens the only real tension is at the borders between territories.

In a sense all plants compete with all other plants. They are all vying for space in which to grow. Studies of plant distributions provide some insights into the way competitive exclusion translates into spatial patterns. One of the key workers in this area, Chris Pielou, carried out extensive studies in Canada into the ways in which plants arrange themselves spatially. One of the most interesting cases is what happens on an *environmental gradient*. This is an area where some physical property changes across the landscape. Examples include temperature declining up the side of a mountain, increasing salinity in a tidal swamp as you move closer to the sea, or decreasing moisture in the soil as you move from a gully up a slope to a ridge. Now, the thing is that each species has a particular range of moisture regime that it prefers. There are absolute limits outside which the plant simply cannot grow at all. To take one extreme, most plants die if submerged in water; otherwise our lakes and streams would be choked with plant life. At the other extreme, plants also die if left in desert soil that is completely dry. They prefer some intermediate range of moisture and some plants require more than others.

At first glance we would expect that, on a hillside (say), each plant species would occupy the full range that it can tolerate. However, this does not normally happen. What Pielou and others found was that species ranges tend to be cut short. Rather than finding (say) fewer and fewer rainforest plants as we move uphill, their distribution finishes, sometimes abruptly, because they cannot compete effectively at the ends of their potential range. This process often gives rise to very distinct zones, such as intertidal rock pools or the bands of different vegetation that we see on the side of a mountain.

So what? What good does it do to know that competition leads to sharp boundaries? There are several immediate implications. One concerns our ability to predict the distributions of species. Earlier we looked at the compilation of biodiversity data. For conservation purposes it is not enough to know that a species is out there somewhere. We need to know where. We need to be able to pinpoint its distribution exactly. For instance, if a new road is going to be built planners need to know whether it will have any impact on rare or endangered species. If we know the climate 'preferred' by a given species then we can predict its potential distribution by looking on a map for all sites in the landscape that fall within that envelope. By 'climatic envelope' we mean the range of values that the species tolerates for rainfall, temperature and so on. So if we know what areas may be home to rare species we can go and check to see whether they are found there. We can also use this approach to check for other areas that might be ideal for species that are under threat elsewhere.

Now, the problem is that competition ruins the simple approach outlined above. Because of competition, the plants and animals that we are concerned about may be completely absent from an area where they could thrive. Competition may also prevent them from becoming established in new habitats. Conversely, if we introduce a new species into an area where it competes with native species there is a risk that it will wipe out the native species altogether.

Is the cane toad an evil force?

Early in the 1930s the Australian sugar crop was under threat. The cane beetle was causing great damage in the sugar cane fields of North Queensland. In an attempt to control this curse the government introduced the cane toad from Hawaii.

Rather than the instant cure that was intended, this action turned into an environmental disaster. Not only did the cane toad eat the cane beetles and their grubs, as intended, it also ate anything else that would fit in its mouth. What is worse, the toad's skin exudes poison, which kills any predator that eats it. The cane toad quickly became a greater pest than the pests it was brought in to control. Today it occupies the entire coastal plain of Queensland as well as northern New South Wales and part of the Northern Territory. There is a very real danger that it might lead to the extinction of a host of native carnivores and other small animals.

Why are introduced species always so damaging to the environment? The fact is that they are not. Most introduced species have to struggle for survival. They cannot compete against the locals. Ask any gardener! The introduced plants and animals that we see as pests are just those species that are highly competitive. Blackberries are not a pest in Europe where they are constrained by competition with many other plants, but in Australia they are a serious weed. Conversely, the eucalypts are severely inhibited within Australia, their homeland, by poor soils and insect attack. In North America, where these restraints do not exist, some eucalypt species grow at incredible rates and are a serious weed.

Interactions between species influence the structure and function of entire ecosystems. An important problem for ecology is to study how ecosystems are 'organised'. One way of looking at this question is to ask how energy and nutrients move through an ecosystem. An important pattern is the *carbon cycle*. Plants draw carbon from the soil and fix it in their tissues. Herbivores eat the plants and absorb the carbon into their bodies. The carbon passes on to predators when they eat the herbivores. When these animals die in their turn, decomposers break down their bodies and return the carbon to the soil, so completing the cycle.

Biologists have spent a lot of time trying to model the ways in which plant and animal populations change through time. They quickly realised that, although foxes and rabbits and spiders and flies may be vastly different animals, the ways in which they interact are fundamentally identical. In each case one animal is a predator and the other its prey. Predation—one species eating another—is just one of many types of interactions that can occur between different animals. There are hosts and parasites, in which the parasite does not kill the host but, like ivy clinging to a tree trunk, benefits from the host in some way.

Cycles and feedback

In general, if there are more species in an ecosystem then there will be more interactions, and more complex interactions. The American ecologist Richard Levins delights in telling stories about the dangers of interfering with rainforests and other complex ecosystems. Removing one innocuous plant could set off a complex chain of events with disastrous consequences. He argued that it was virtually impossible to model a rainforest accurately. There are simply too many variables and too many interactions to measure. Instead, he suggested a very simple, yet effective, way to study processes in these complex communities.

Levins's *loop analysis* looked at the patterns of interactions between different species. We can simplify the many kinds of possible interactions between species by asking whether they are positive or negative for each of the species concerned. A positive interaction is one that tends to help the population increase in size. A negative interaction is one that decreases the size of the population. A host species has a positive effect on a parasite. The more hosts the more parasites that can live. A predator has a negative effect on its prey—the more predators there are the faster the prey animals will die.

The power of this way of looking at nature becomes apparent when we consider whole chains of interactions. Take foxes and rabbits. The FOX–RABBIT interaction is negative: more foxes kill rabbits faster and lead to fewer rabbits. The RABBIT–FOX interaction is positive: more rabbits mean food for more foxes, so the fox population can increase in size. This combination of effects leads to a feedback loop. It is a loop because we can trace a series of interactions from rabbits to foxes and back to rabbits again. It is negative because the sequence contains an odd number of negative interactions.

So what does this mean? What happens if we put the foxes and rabbits together? Well, as the foxes start killing rabbits, so the rabbit population falls. But fewer rabbits means that there is less food for the foxes, so in time the fox population starts to fall. This decrease in fox numbers means that more rabbits can survive, so the rabbit population increases. But that means that fox numbers can increase again, and so on. So over time the whole system cycles, with the numbers of foxes and rabbits oscillating up and down.

This kind of system is known as negative feedback. It is called feedback because the effects of any change in one of the components eventually 'feeds back' to affect that component itself. In negative feedback, the system tends to cancel out any change in the components. So any increase in the rabbit population produces effects (namely more foxes) that tends to decrease rabbit numbers again.

Feedback is a well-known process. Engineers have studied and exploited feedback for many years. A speed limiter on a car, for instance, is a negative feedback device that is designed to counter any tendency of the vehicle to go faster than a set speed. The other type of feedback loop is positive feedback. In positive feedback the net effect of the loop is positive. This means that any change tends to be self-perpetuating. That is, the change keeps on going, and usually accelerates.

In general, negative feedback promotes stability in a system—it tends to negate change. A host of man-made devices, from speed delimiters to temperature regulators, exploit this principle. Positive feedback, on the other hand, promotes instability. It feeds on any change introduced into the system, and leads to further change. Now, in any ecosystem there will be lots of interactions between the different populations that make up the system. The question is whether those interactions will produce positive or negative feedback.

Positive feedback in an ecosystem is disastrous. Suppose that positive feedback starts affecting some

Figure 4.1 Coupled feedback loops from the Limits to Growth simulation model of world population (see Chapter 3). The arrows indicate the kind of effect (positive or negative) that different variables have on each other. We can determine the nature of each loop by counting the number of plus and minus signs within it. An odd number of minus signs shows that the loop is negative; otherwise it is a positive loop. In a negative feedback loop any increase in a variable will 'feed back' to it and tend to decrease it again.

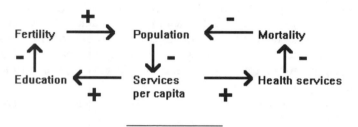

animal population. Then each change in the population will lead to further change in the same direction. If the population is falling then it will keep falling—at an ever-increasing rate—until the population crashes to zero. This is what stability theory predicts for competition. If two species (foxes and weasels, say) are competing then the effect of each population on the other is negative. Although it is not at first apparent, this situation produces positive feedback. More foxes means fewer weasels, and fewer weasels in turn means more foxes. So the system tends to magnify any change. So such a system is inherently unbalanced. Any increase in one population will perpetuate further increases, and simultaneously drive the competitor to extinction.

Now let's look at what actually happens in a large community containing many plant and animal species. Naturally there will be many interactions, of many different kinds, between the populations involved. These interactions will produce many feedback loops. Suppose that we could select the species at random. Then we would expect that about half of the interactions would be negative and about half would be positive. So the resulting system should contain both negative and positive feedback loops. Those positive feedback loops have the potential to disrupt the entire system.

In the early 1970s the Australian physicist Robert May caused a sensation in the world of ecology. In the course of many years of observing natural communities, ecologists had come up with various generalisations. One of the best known of these rules was that complex ecosystems tend to be more stable than simple ones. This principle was based on the observation that complex communities, like rainforests or coral reefs, tend to recover faster than simple ecosystems. The thought was that complexity somehow enables a system to resist change. May caused a shock by showing that in artificial systems the exact opposite is true. The more complex a system is the more likely it is to be unstable.

We can appreciate the truth of May's argument at once by thinking about feedback loops. As we saw above, systems with lots of species and lots of interactions are likely to be destabilised by positive feedback loops. Experiments confirm this intuitive conclusion. Now, it is simply not possible to carry out experiments on this matter in the real world. They would be too slow, too costly, too difficult and too disruptive. Instead we can carry out experiments on a computer. To do this we make up model communities and have the computer assign random interactions between the different species. We can then run the model to see what happens to the mixture of species that we have formed. Not surprisingly, most of the models contain positive feedback loops and collapse. That is, one or more species goes extinct in the model world that we have created. Only a tiny fraction of the model communities retain all of their species.

Experiments like these provide important insights about the nature of ecosystems. Perhaps the sort of collapse that we see in the models goes on in nature all the time. Perhaps the ecosystems that we see around us are just the combinations that do not collapse. Or perhaps systems are collapsing all the time, but we are just not aware of it. In any case, ever since May raised the question about stability and complexity, ecologists have been trying to understand whether complexity is associated with stability in ecosystems, and if so why.

The issue is not at all simple. For a start we have to ask many questions about the model ecosystems. For instance, we saw earlier that competitive exclusion does not always occur in landscapes because the competing populations are spread over the landscape and may not always be interacting directly. Other factors come into play as well. It is one thing to say that a certain set of species forms a positive feedback loop, but how strong are the interactions compared to other processes that are going on at the same time? They may be very weak. The above effects are especially important when one population is involved in several different loops.

Time delays have an enormous bearing as well. Sure, more rabbits may make life easier for the fox population, but it takes time for the foxes to breed and for their offspring to reach an age where they can start hunting rabbits. Delays of this kind tend to produce oscillations of the kind we saw before.

Despite the uncertainties that remain about the role of feedback loops in ecosystems, they do provide several strong messages. One is to emphasise just how dangerous it is to introduce new species into an environment. As with the cane toad, the history of conservation is full of stories about well-intended introductions of exotic species leading to environmental disasters.

IS THERE A BALANCE OF NATURE?

Popular portrayals of ecology always emphasise the idea of balance. They are full of expressions such as 'the natural cycle', or 'upset the balance of nature'. The idea of equilibrium—balance—is almost dogma in some sections of the environmental movement. And yet is it really true? Is there a balance of nature? Does such a thing really exist? The above discussion has already raised serious questions about the real stability of ecosystems. Ecologists essentially abandoned assumptions about equilibrium years ago. To some extent our perception of stability and balance in the natural world is an illusion. It is worth looking briefly at why this is so.

In the south-east of the Australian state of Victoria lies a vast wilderness of hills covered in forest. When one of the authors was slightly younger he used to spend a lot of time bushwalking and camping in these forests. Go out there today and the region still looks just as it did all those years ago. There are tall trees and mossy rocks and rainforest gullies clustered by the creek beds. And yet there are signs that something has changed. Poking up above the canopy in many places you can see the skeletons of enormous trees. Those skeletons were there when the author was young. They are still there today. Taken at face value, one could almost believe that they had always been there. But the truth is that they are the remains of trees killed in an enormous bushfire that ravaged the region in the year 1939. The size of these skeletons sticking above the canopy shows that the forest we see today has still not reached maturity. Although it looks stable and constant, certainly throughout one person's lifetime, those

81

stumps are a vivid reminder that great changes have taken place in the past.

The pace of change in ecosystems varies enormously. Populations of fruit flies can turn over in a few days. On the other hand we can find in California and elsewhere groves of trees that were growing before Christ was born. Events in some ecosystems, especially forests, can take so long that they look as though no change is happening at all. In short, they *look* stable. It is the time scale that deceives us.

CHAPTER 5

THE ABC OF NATURE

Philosophy is written in this grand book—I mean the universe—which stands continually open to our gaze, but it cannot be understood unless one first learns to comprehend the language and interpret the characters in which it is written. It is written in the language of mathematics . . .

—Galileo in *Il Saggiatore*, 1623

. . . like a child, we can spell out the alphabet without understanding more than a few words on a page.

—Nobel Prize winner William Gilbert, on the genetic code.

Every age has its own way of viewing the world. During the industrial revolution the world was seen as a great machine. This view coloured people's thinking about all sorts of matters. The heart is a pump; the arms and legs are bio-mechanical levers. The solar system operates like clockwork.

Today we are in the midst of an information revolution. Telephones, TVs and computers are everywhere. They dominate the lives of many of us. Instead of machines it is now computers that colour our view of the world. The computer seems to be a model for understanding many processes in the natural world. For convenience we'll call this view the 'nature as computing' paradigm. Perhaps the most seminal image is that of the genetic code as life's program. In a literal sense there is a striking resemblance between the sequence of bases in a strand of DNA and a computer program written on a magnetic tape. But the similarity goes much deeper still. Like a magnetic tape, a DNA strand can be copied to produce new DNA, or to produce RNA which is similar but has only a single strand. The body uses RNA as a template for manufacturing proteins. The sequence of bases along the DNA chain forms the genetic code. DNA has four bases: Adenine (A), Cytosine (C), Guanine (G) and Thymine (T). RNA substitutes Uracil for Thymine. Because DNA consists of paired chains, it always contains two sequences that are complementary to one another. So, for example, the sequence AATGCTG is complementary to the sequence TTACGAC.

Two crucial processes occur when the DNA helix unwinds and tears apart. One is that each half of the chain can form the template for a complete new chain. The other is that the chain can be read to provide the data for growth and other processes. This happens when the code is copied to form RNA, which is another type of chain. RNA is a single strand copy of the DNA sequence. However, instead of Thymine it contains Uracil as the fourth base.

The crucial process is the interaction of mRNA with ribosomes to manufacture proteins. Ribosomes are almost literally small molecular machines. They attach themselves to mRNA and move along the chain. Each site they attach to serves as a template for forming amino acids. As they come off the production line, the amino acids string together to form proteins. In this process the ribosome acts very much like a tiny Turing Machine (see Chapter 2), reading in the RNA like a data tape and outputting sequences of amino acids to form proteins. Several aspects of the above process are worth noting. First, as a template for amino acid production, the ribosomes use triplets of bases (called *codons*). So the codons are triplets, such as AAT or GCA. Each codon corresponds to a single amino acid. Now there are four bases at each of three positions within a codon, so there are 64 (=4³) possible codons. However, there are only 20 amino acids, so there is a lot of redundancy. In most cases the third base in a codon is irrelevant; the first two bases determine what the resulting amino acid will be. This has led to speculation that the present code is derived from an earlier form in which codons consisted of pairs of bases.

Another computer-like feature of protein production is that, just like a computer program, there is a start codon, which tells the ribosome where to start making amino acids, and codons that tell it where to stop. Notice that because codons consist of three bases each, there are three possible places where codons can start. Also there are two possible directions. So the same piece of code can be read in six possible ways, called *reading frames*. An *open reading frame* is a segment of code between a start and a stop codon. A gene is a segment of DNA that codes for a particular function. Usually this function is to produce a protein. However, genes can also have several other roles, such as controlling the operation of other genes. An interesting side effect of having several different reading frames is that the code for different genes can overlap. Such overlaps are possible because the redundancy of the code allows some freedom for bases (especially the third position) to change (silent mutations) without altering the resulting amino acid sequence. This property of overlap is most frequent in organisms, such as viruses, that have relatively little DNA.

Another strange feature of DNA is the enormous amount of *junk DNA*. Only a fraction of the DNA strand actually contains genetic code. The rest (as much as 80 per cent) contains base sequences that apparently are never read. This fact posed a great puzzle for molecular biologists. Why should organisms waste so much 'effort' on building junk DNA? Remember that DNA is present inside every cell in the body. So the amount of food and energy tied up in the production of junk DNA is considerable. It's just not like living things to be as inefficient as these figures suggest. There must be some advantage in having so much spare DNA lying around.

Perhaps the most likely answer is that junk DNA is needed to enable the reading of gene sequences. Although its primary structure is a helix, the DNA molecule also tends to twist itself up into complex knots—the so-called secondary and tertiary structure. So most of the sequence may be inaccessible for most of the time. The coils of DNA need to unravel at least slightly for copying to take place. The junk DNA may help to ensure that all of the coding regions are accessible during this process.

Another suggestion is that the junk DNA provides space for writing new genes. Overlapping genes do occur (especially in viruses). But although overlap represents efficient storage, it also inhibits variation. Mutations in one sequence would be likely to affect an overlapping sequence as well. Perhaps both of these explanations (not to mention others) are at least partly valid.

VIRTUAL PLANTS

Computer-like processes in nature are not restricted to protein synthesis and development. In particular, the idea of language in nature goes well beyond the coding of our genes. Biologists have turned increasingly to language as a means of describing the ways in which biological systems are organised.

When we say 'language' here we don't mean everyday speech. The problem is that normal language is altogether too flexible. Instead, we use artificial languages that are built specially to represent particular systems. These *formal languages* are just rules for manipulating symbols. Traditional mathematics, with all its symbols and equations, is really a formal language.

To appreciate the nature of formal languages, and the way they differ from normal languages, we need to go back to grammar school briefly:

<sentence>	→	<subject> <verb> <predicate>
<subject>	→	<article> <noun>
<predicate>	→	<object>
<predicate>	→	<adverb phrase>
<adverb phrase>	→	<preposition> <article> <noun>
<preposition>	→	on
<object>	→	<article> <noun>
<article>	→	the
<article>	→	a
<noun>	→	cat
<noun>	→	rat
<noun>	→	cheese
<noun>	→	mat
<verb>	→	sat
<verb>	→	ate
<verb>	→	chased

The above rules (a *grammar*) define a formal language. This language is simply the set of all sentences that can be derived by applying the above rules. Notice that the rules contain two sorts of items: terms in brackets (*grammatical variables*) and terms without brackets (*grammatical constants*). The derivation process consists of starting from the grammatical variable <sentence> and using the rules to replace each variable until there are none left. Here is how we would derive a familiar sentence:

```
<sentence>
<subject>                 <verb> <predicate>
<subject>                 sat    <predicate>
<article>  <noun> sat            <predicate>
<article>  <noun> sat            <adverb phrase>
the        <noun> sat            <adverb phrase>
the        cat    sat            <adverb phrase>
the        cat    sat            <preposition>    <article> <noun>
the        cat    sat            on               <article> <noun>
the        cat    sat            on               the       <noun>
the        cat    sat            on               the       mat
```

The richness of the language arises from the number of choices that we have for replacing each variable as we go. Here are some other sentences that we can derive from the above grammar. See whether you can work out how they are derived:

the cat chased a rat
a rat ate the cheese
a cheese sat on the mat

Do not be fooled. This language is *not* English! It may look like English. In fact it is a subset of English. But there are many sentences we can make up within this language that make little or no sense at all. For example:

a mat chased the mat
the cheese sat the rat

Also the language can be very restrictive. We can make up many perfectly sensible English sentences, using the words given, that are not part of the above language. For example:

the rat ate cheese

L-systems

Have you ever looked closely at a plant? Next time you're in the garden look at the pattern of branches, leaves and flowers on a bush or a small tree. The arrangement is usually very regular (at least before damage by insects, storms and wilting). L-systems allow us to capture that regularity in a computer model.

Biologists have always been struck by the symmetry and regularity of plant growth. In the late 1960s the biologist Aristid Lindenmayer realised that formal languages could capture these patterns. Borrowing the first letter of his name, the approach that he pioneered is now called L-systems.

Figure 5.1 An L-system model of a branch. This picture was generated using the freeware program FRACTINT.

L-systems apply the above grammatical approach in order to model growth. Instead of terms about cats, rats and prepositions, they use terms that refer to (say) parts of a plant.

To understand this, let's take a simple example, defined by the following rules:

Variables :   A, B
Constants :   none
Start :   A
Rules: A → B
      :   B → AB

This L-system produces the following sequence of strings:

Stage 0: A
Stage 1: B
Stage 2: AB
Stage 3: BAB
Stage 4: ABBAB
Stage 5: BABABBAB
Stage 6: ABBABBABABBAB
Stage 7: BABABBABABBABBABABBAB

Counting the length of each string yields the famous Fibonacci sequence: 1 1 2 3 5 8 13 21 34 . . . . This simple sequence also demonstrates the way in which iterative properties of L-systems lead to self-similarity and hence fractal growth patterns.

Growth patterns for two tree species

Simplified models are given beneath the diagrams. Names in angle brackets denote growth structures and arrows denote growth processes. The states of particular struc-

Figure 5.2 Branching patterns in *Eucalyptus alpina* **and** *Hakea salicifolia.*
The rules given in the text describe the organisation.

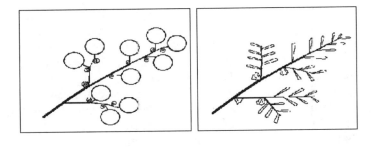

tures are indicated by colons. The models differ syntactically in the timing of fruit production and semantically in the shapes of leaves and fruit.

(a) *Eucalyptus alpina*:

```
<bud: leaf, angle>       → <meristem> <bud: stem, angle>
<bud: stem, angle>       → <stem> <bud: leaf, angle + increment>
<meristem>               → <leaf> <fruit> <bud: juvenile, 0>
<bud: juvenile, angle>   → <bud: stem, angle> /delay=1 year
```

(b) *Hakea salicifolia*:

```
<bud: leaf, angle>       → <meristem> <bud: stem, angle>
<bud: stem, angle>       → <stem> <bud: leaf, angle + increment>
<meristem>               → <leaf> <fruit> <bud: juvenile, 0>
<bud: juvenile, angle>   → <bud: stem, angle> /delay=1 year
```

VIRTUAL ANTS

Another popular image in computing is that of the robot. Mention the word robot and most people immediately think of a mechanical man. However, biologists are more

91

and more aware that the behaviour of many organisms is like that of a robot. Insects, for instance, with their simple repertoire of stimulus–response actions, behave as though they are programmed to respond in a fixed way. One of the really profound differences that computers have made is that they allow us to go beyond numbers and calculations. Computers originated as calculating machines. Charles Babbage designed his 'Analytic Engine' as a tool for generating mathematical tables. But the origins of computers also included ideas and methods that were not about calculation. The idea of a computer program, for instance, originated with punched cards. The first punched cards were introduced to control the fabric patterns produced by textile looms. The first electronic computers, built during World War II, were used to crack secret codes. Computer scientists were very quick to realise that they could use computers for all sorts of manipulations.

The links between computers and language go back to the beginnings of computing. As we saw in the first chapter, computer languages arose from the need to develop complicated procedures that can be repeated at will.

In principle there is no real difference between a computer program that carries out a numerical calculation and one that manipulates text. One involves moving and copying numbers; the other involves moving and copying words. For instance, adding up a shopping list involves taking numbers one at a time from the list and adding them to a running total. Searching for a given term in a list of names involves checking each item in a list for a given value or string pattern. One of the most profound insights of the 'nature as computing' paradigm is that scientific models of nature need not be restricted to mathematical formulae. For example, just as we can use computer languages to write computer programs, so we can use languages to describe processes in nature. To illustrate this, let's look at turtles before we go on to ants.

The turtle as artist

In the late 1970s the Stanford mathematician Seymour Papert devised a system of graphics called *turtle geometry*. The aim was to make it easier for children to learn mathematical ideas by embodying them in concrete form. The basic idea of turtle geometry is to think of a geometric shape as the trail left by an imaginary turtle as it wanders around on a surface. Suppose that we attach a pen to a turtle and place it on a table covered with paper. Then, as the turtle moves around, the pen will leave a mark that shows where the turtle has been. Now, as it moves, the turtle alternately moves forward (or back) some distance and occasionally turns to the right or left. We can at once devise a simple language to describe what happens. To start simply, suppose for a moment that every turn the turtle makes is a right angle. We can use F to denote one step forward, and L and R to denote left and right turns. Then we can describe a trail in terms of the sequence of actions that the turtle makes as it moves, such as FFRFFFLFRFF. Strings of symbols like this form a simple language. Not only can this language describe a path, it also allows us to tell the turtle how to draw a pattern. In the computer programs that implement turtle geometry, the 'turtle' is usually represented by a dot moving around on the computer screen. This dot moves around drawing a pattern in response to any string of actions that we present to it. For example, the string FRFRFRFR tells the turtle to draw a square.

Turtle geometry embodies many mathematical ideas. The idea of an inverse, for instance, is widespread in mathematics: in arithmetic, subtraction is the inverse of addition; in calculus, differentiation is the inverse of integration. Turtle geometry embodies the notion of inverse in a very visible way. For every action that the turtle can take there is an inverse action that returns it to its previous state. The inverse of F (go forward) is B (go back); the inverse of L (turn left) is R (turn right). So the string FB or the string LR each indicates that the turtle immediately undoes the action it takes first. The idea of an inverse extends to any path at all. To obtain the inverse of a given path, we simply reverse the string and replace each term with its inverse. Here is how this procedure works for the example given earlier:

The original path:	FFRFFFLFRFF
Reverse the string:	FFRFLFFFRFF
Replace each term:	BBLBRBBBLBB

So the path FFRFFFLFRFFBBLBRBBBLBB would see the turtle wander a short way and then backtrack to its starting point.

Turtle geometry also embodies many computing ideas. For instance, if we give names to particular paths, we can use those names to denote whole elements within a larger pattern. In drawing (say) the square as described above, we could give the string FR the name 'SIDE'. The whole pattern FRFRFRFR then reduces to '4 SIDE'. However, we can go a lot further than this. If we then give the whole pattern the name 'SQUARE', we can use it as an element in drawing bigger and more complex patterns. Using names for patterns in this way is exactly like giving names to subroutines or functions within a computer program.

Turtle geometry embodies iteration too. For example, we can allow the turtle to make turns at any angle, rather than always 90°. Suppose that we modify the element

Figure 5.3 Patterns generated by turtle geometry. The paths shown are described in the text.

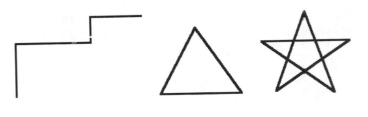

SIDE above to be SIDE(ANGLE), where ANGLE is a number telling the turtle to turn by whatever number of degrees. That is, SIDE(ANGLE) is another name for the string 'F R(ANGLE)', where 'R(ANGLE)' means 'turn right by ANGLE degrees'. If we set ANGLE to be 120 and set the turtle walking, then after drawing three sides its trail starts retracing itself. So the resulting pattern is a triangle. If we repeat this experiment with the value of ANGLE set to 72, then the result is a pentagon. Other angles produce more unusual figures. With ANGLE set at 144, the result is a five-pointed star. With ANGLE set at 95, the result is a rosette pattern.

The point of the above discussion is that turtle geometry is intimately tied to the turtle language. Each expression in the language corresponds to a pattern on the surface. By our introducing more elaborate rules into the language, turtle geometry becomes capable of creating virtually any pattern at all. Let us look at two simple examples. First, we can introduce a growth operator, G, that magically shrinks or enlarges our imaginary turtle. So the command 'SIDE(ANGLE) G(0.9)' would draw a side and then shrink the turtle to 90 per cent of its original size. If we repeat this process, then the result will usually be an inward spiral. If we set the growth factor greater than 1, then the path will spiral outwards. Finally, we can greatly enlarge the ability of the turtle language to describe complex patterns by exploiting the ability to give names to strings. As an example, look at the following set of definitions:

```
SIDE(ANGLE)  =  F R(ANGLE) ·
CELL         =  6 SIDE(60)
CLUSTER      =  6 (CELL F L(60))
HONEYCOMB    =  CLUSTER SIDE(60)
```

In this set the CELL produces a hexagon, CLUSTER produces a cluster of hexagons and HONEYCOMB (when repeated) produces a honeycomb pattern. This final example embodies another important idea in computing: modularity. If we were to sit down and try to write out a string to describe the turtle's complete path in drawing the honeycomb pattern, it would be a long and difficult process. However, by breaking it into a hierarchy of modules that produce different levels and elements of the pattern we can approach the problem in a simple and systematic way.

From turtles to foraging ants

How does turtle geometry relate to the behaviour of real animals? Perhaps, instead of turtles, we should look at ants. To move from the simple 'behaviour' embodied in turtle geometry to the behaviour of real animals requires several changes. First we have to consider how real animals move. All of the patterns set out above were deterministic. That is, the path taken was completely predictable from the start. A foraging ant, on the other hand, moves about essentially at random. It does not move according to a fixed pattern. Secondly, and most important of all, an ant interacts with its environment.

Many of its movements are determined by what it encounters as it moves around. If an obstacle blocks its path then it searches for a way around. If danger threatens then it goes into panic mode and tries to escape. If it detects food then it homes in on the scent. Finally, ant behaviour is not solely concerned with movement, but with all manner of activities. Many of the rules that govern an ant's behaviour concern the type of behaviour, rather than the exact details of how it moves.

Another natural phenomenon that is literally computing is the processing of sensory information. Just like a computer, our senses are the input devices of our brain; speech and action provide the outputs. Sensory perception is a matter of selectively throwing away information. Take human vision, for instance. Every second our eyes receive more information than we can possibly process in detail. The miracle of vision is that we are able to distil literally terabytes of sensory data into a manageable volume of information. The process of sight thus involves selective ignoring of most of the details transmitted, so that only vital features remain. What interests us in the world are the shapes and reflectance of surfaces. We're not interested in absolute brightness level or even the actual spectral distribution of light that comes into the eye.

It's a common experience that colours under fluorescent light may look different outside. This is an example of a breakdown of a visual strategy which normally works very well: colour constancy. If the spectral distribution of the light changes (e.g. at sunset, when it shifts to longer wavelengths, or the red end of the spectrum) then the light reflected from a surface changes. But that has nothing to do with the actual surface and it would confuse the recognition process. So we work back from the spectral distribution of light in the early stages of vision to *colour* in a module of the brain referred to as area V4.[9] We've thrown away the specific details of the light's brightness and spectral distribution.

Similarly, snakes, for example, can have trouble seeing freshly killed prey right in front of them and yet are able to detect even the slightest movements in their field of view—their eyes are adapted to select the sorts of information that are vital to their survival. Snakes want to kill live prey rather than eat carrion and can therefore ignore a lot of static pattern information used by other animals. This ensures, of course, that the meat is fresh but in some snakes it serves another function. Snake venom has pre-digestion enzymes that can be usefully circulated in the animal's blood stream before it dies, making the bite from some snakes, such as the African puff adder, particularly nasty even after hospital treatment has avoided death.

How to get up in the morning

When we look at human behaviour, there is good evidence from psychology that humans behave a bit like robots. In particular, we all develop habits. Habits are fixed patterns of action that we follow in a given situation. For instance, one person's habit might be to scratch his or her head while thinking; someone else might bite the end of a pencil while writing. On a higher level, regular activities such as going for a drink after work may be habits too. When we think of habits we tend to think of unusual, and often undesirable, behaviour—such as a child picking its nose. However, habit-formation seems to be linked to important processes in human learning and behaviour. The great French child psychologist Jean Piaget introduced the idea of schemas to help explain mental development in children. Piaget suggested that children learn by developing what he called 'schema', or patterns of recognition, thought and action. A schema is a 'recipe' for dealing with a particular type of situation. For example, if you walk into a room for the first time, you expect to see certain features: walls, ceiling, floor, furniture and so on. The importance of identifying the features that we expect is that once we have identified them we can forget most of the details. In general we need remember only the features that are unique or distinctive. A room with no floor, for instance, would certainly be memorable. As we saw earlier, this selective forgetting assists perception, because we do not need to process so much information.[10]

The above expectations, and the way you react to them, constitute what might be termed an 'entering a room' schema. An important mode of learning is what Piaget called 'assimilation'. We assimilate (i.e. cope with) many new experiences by adapting existing schemas that deal with similar situations. For example, the idea of entering a room generalises to entering all kinds of structures, such as a shed, a tent or a cupboard. Genuinely new experiences (e.g. travelling overseas for the first time) are memorable because we do not have the perceptual framework set up to help us ignore details.

And if you still don't believe that humans are programmed automata, then consider the following question. What happens when you get up and go to work each morning? Most people answer this question with a list something like the following:

get out of bed
have a shower
get dressed
have breakfast
pack my briefcase
leave home
catch the bus

99

The details may vary from person to person, and even from day to day. But we can all identify with this sequence of events. Most of us have a routine that we follow on a regular basis. The point is that we are all creatures of habit. The above sequence is essentially a program for getting up and going to work. What is more, each step, each 'routine' in our going-to-work program, has subroutines. Take just one step, say having breakfast. A typical 'subroutine' for having breakfast might be as follows:

    get out cereal, milk and spoon
    fill bowl with cereal and milk
    heat bowl in microwave
    sit and eat cereal
    clear breakfast table

Each of these steps can again be broken down into a yet finer repertoire of habitual routines that we use. Eating cereal, for instance, might follow a sequence something like this:

    dip spoon in bowl
    remove spoonful of cereal and milk
    put spoon in mouth
    chew and swallow cereal
    if bowl contains more cereal
    then repeat the above steps
    or else finish eating

The point of all this is that our behaviour is built up out of schemas piled on schemas. Elementary schemas, such as holding a spoon, are learned in early childhood. In later life we exploit them as givens from which to create more complex behaviour patterns. In the following sections we shall see how new features can emerge out of simple rules of behaviour.

From syntax to synergy

Suppose that you are caught in a blizzard and know that there is a hut at the top of the mountain. How would you find the top? The answer is simple. Just keep heading uphill until you reach the top. This example is a simple case of a 'rule of thumb'. We use such rules constantly in our daily lives. Many well-known sayings are really rules of thumb: 'Never stand under a tree during a thunderstorm'; 'Do unto others as you would have them do unto you'. Likewise, many rules—e.g. 'Keep well away from other cars when driving'—are learned from experience. Rules of thumb are ways of coping with uncertainty. Given any problem to solve, and a complete set of facts, we can usually nut out the best solution. For instance, suppose that you are driving through a strange city; then if you have a street map you can work out the route that will get you to where you want to go. Unfortunately, life is rarely that easy. What do you do if you are trying to find your way through a strange city without a map? Well, of course, you use a simple rule, such as to follow major roads heading in the direction you want to go. The point is that, in the absence of complete facts, we cannot plan a strategy that we can guarantee will 'win' (achieve our desired goal). We may find ourselves caught on a freeway with no exits. We just don't know in advance what will work and what will not. So we adopt a strategy that experience shows will work most of the time.

Next we see that rules of thumb can have profound consequences when applied *en masse*.

The bull ant and the bumblebee

The walk home from our university passes through several hectares of bushland. At one point there is a clearing where a road passes through the trees. By the side of this road there is a large ant nest, one of many in the area. The nest is easy to spot: it is marked by a circle of cleared earth about ten metres in diameter. Late in the afternoon the ants are very active. Thousands of them swarm all over the nest. Every shred of litter is at once cleared away by workers. Meanwhile, soldier ants guard two dozen or more entrances. Thousands more ants head off into the bush to forage for food.

The nest gives every sign of being well organised and well designed. If we could look inside, this impression would be even stronger. Eggs and food are stored in separate areas and the nest as a whole is kept clean and tidy.

Ants are not alone in building elaborate homes for themselves. The precision and regularity of a beehive, for example, is even more striking.

How do they do it? That animals with no real intelligence can design and build such well-organised nests has long been a source of wonder.

The answer to this puzzle is simple, but amazing nonetheless. There is no blueprint, no central planning. Ants and bees have no idea what a complete nest should look like. The insects just act according to a very simple repertoire of behaviour. An ant nest or a beehive emerges from myriad simple interactions of individuals with each other and with their surroundings.

To appreciate how an orderly nest can emerge in the absence of an overall design, let's look at how ants sort materials. Suppose that an ant is wandering around the nest and comes across a stray egg. It picks up the egg and carries it away. It has no plan that tells it where to take the egg; it simply carries the egg with it as it wanders

Figure 5.4 **Formation of a virtual ant colony in the computer program Xantfarm. Instead of grains of sand the ants move screen pixels, creating hills and tunnels.**

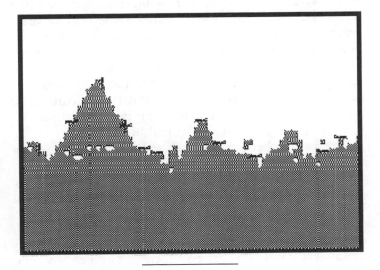

around the nest. Eventually, in the course of its wandering, it comes across a pile of eggs and drops the egg that it is carrying. Now this pile of eggs may start off simply as a random concentration of two or three eggs. However, as the ants add more and more eggs to the pile, it grows into the nest's central egg repository. The whole process requires only two rules of behaviour to make it work:

- IF you see an egg THEN pick it up
- IF you see a pile of eggs THEN drop the egg you are carrying

In Chapter 2 we saw how foraging and other aspects of ants' social behaviour can be described by equally simple rules.

103

There is an amusing computer program, Xantfarm, that embodies the above ideas.[11] Xantfarm is used as a screen background on Unix workstations. To represent the ground there is a line across the screen, with white sky above and green earth below. Little 'virtual ants' wander back and forth on the ground, as well as up the sides and across the top of the screen. Whenever an ant finds a loose pixel (a virtual grain of sand) it picks it up and deposits it on a pile of sand. Leave the program running for several days and the virtual ants create huge piles of sand with tunnels burrowing into them from the side. In short, they build a virtual ant colony.

Interaction between 'agents' and their environment is an important source of order in many natural systems. Some of the key early work on this topic was done by the team of Paulien Hogeweg and Bruce Hesper at the University of Utrecht. They showed that the structure of many living systems, such as ants or bumblebees, could be explained by the TO DO principle. That is, the insects simply do whatever there is to do at any time and place.

The TO DO principle has many potential applications. Take robotics, for instance. The traditional approach to building robots was to pass a robot's visual input to a large computer, analyse a scene, and then make decisions about action. For years this centralised approach became bogged down. In effect, robots were trying to understand their entire environment before they could act. However, Rodney Brooks, a researcher at MIT, showed that such elaborate interpretation was not necessary.

In his robots, legs and other systems are all independently controlled. Simple rules govern how each leg responds to circumstances at any given time. If a leg is tilted back, for instance, it is programmed to lift and move forward. Feedback systems ensure coordination between the legs. Behaviour of the entire robot emerges out of interactions between all these different systems. No central planning or intelligence is involved.

One result is that robots can now be made smaller and more adaptable than ever before. Some of these robots resemble large insects, not just in their appearance but also in their behaviour. In the future a home may come to resemble an ant colony. Imagine a hundred tiny robots wandering around the house while you are away, cleaning up, killing cockroaches and sorting belongings.

CHAPTER 6

ON THE IMPORTANCE OF BEING WELL CONNECTED

She has no money, no connections, nothing that can tempt him to—she is lost for ever.

The upstart pretensions of a young woman without family, connections, or fortune. Is this to be endured! But it must not, shall not be.

Jane Austen, *Pride and Prejudice*

In Victorian England, social connections were everything. For any young lady or gentleman in polite society it was essential to be born into the right family, to go to the right school and to join the right club. With good family and social connections came invitations and introductions, good marriages for young ladies and shining careers for young gentlemen. Lacking good connections was a severe handicap when you set out to make your way in the world. Victorian England was not the only society in which social connections were important. They are important in any society. They are still important today.

Just what do we mean by *connections*? Connections are links between people. Besides social links and family links, connections can include just about any way in which people interact or relate to one another. For instance, there are neighbours. There are professional colleagues, clients, customers. There are travellers thrown together by chance on the same train or plane. There are patients in the same hospital ward. Students who study together, people in the same theatre. There are members of clubs, societies, political parties or religions. Many features of society rely on one or another of these connections. In a small town or neighbourhood, for instance, gossip depends on people talking to each other. The spread of disease depends on people coming into close contact with one another. Fads and fashions depend on people influencing each other's likes and dislikes.

THE BALL AND STICK MODEL

At first sight this idea of connections may seem too trivial to be taken seriously. Many scientists have expressed this view. However, the very idea of connections leads to some deep and fundamental insights. It turns out that many ideas that have formerly been studied separately are really consequences of connections. In many respects it is the connections themselves, rather than features peculiar to a specific system, that control the structure or behaviour of a system

Before going any further we need a way to express the idea of connections. One simple approach is the so-called *ball and stick* model. We represent any system as balls (the objects) joined together by sticks (the connections). Now, real people don't look like balls—but remember that this is just a convenience. All we need to capture is the essence of connections. To achieve this it is enough to represent people as balls, and social connections as sticks. In the ball and stick model people and their social behaviour are no different from (say) towns joined by roads or stars tugging each other through gravity. The details of the system do not matter. Only the number and pattern of connections matter.

Mathematicians call the ball and stick model a *graph*. A graph is simply a set of nodes joined by edges. That is, in a graph each ball is called a *node*; each stick is called an *edge*. The ball and stick model is ingrained in our thinking. We draw boxes and arrows on the backs of envelopes to explain many things. Who has never tried to explain, say, the organisation of their company, or their family tree, by drawing names joined by arrows? We also draw electronic circuits and computer flow diagrams as lines joining shapes.

In a very real sense the ball and stick model pervades our thinking. It is even built into our language. Take the sentence:

The dog ate the bone.

If we look at the syntax, we see that this statement takes the form:

subject/verb/object

Both the subject and the object are things (*balls*) and the verb defines a relationship (*stick*) between them. Whether we realise it or not, a lot of science involves identifying ball and stick relationships. For instance, taxonomy is not just about identifying species. It is mostly about identifying how different species (the *balls*) are related (the *sticks*). Many studies are about relationships, such as: *does smoking cause cancer?*

As we have seen, one important insight of the ball and stick model is that many different systems have a deep similarity. The organisation of large corporations has its ball and stick structure in common with the taxonomy of the mammals.

A game of chess

It is beginning to seem as if the ball and stick model applies almost everywhere. And it does. Not only does it refer to how things are put together; it is also relevant to the way they behave, change or grow.

Take a simple example. Suppose that two people are playing chess. We can describe the state of the game at any time by giving the location of each piece on the board. From any given state of play the rules of chess allow for many possible moves. At the very beginning of a game white can choose, for instance, to move any one of the eight pawns or either of the two knights. There are many ways in which white might move a single piece. The king, for instance, might be jumped to the centre of the board. But the rules of chess outlaw this move. Of the 512 ways to move a single piece from the opening state, only 20 are legal.

Now, the important thing to note here is that any move we make connects one state of the board to some other state. Suppose that we could plot all possible states of the chessboard on a map, with each state, each arrangement of the board, represented by a dot. Then we could view an entire game as a path leading from the opening state to one of many possible final states.

When we look up at the sky, or look at a beach, we can at once see how vast are the numbers of stars, or the numbers of grains of sand. Though not immediately obvious, the number of different ways in which a game of chess can proceed is equally vast. To see this let's consider (say) the first ten moves in a game. For the sake of simplicity, suppose that no pieces are taken and that on each move there are 20 possibilities—as on the first move. Then in 10 moves there is a total of 20^{10} or 10^{13} possible sequences of states (i.e. 1 followed by 13 zeroes). However, the paths allowed by the rules of

109

chess are only a small subset of all possible paths within the *state space* of all possible states of the board. If we allow any move of one piece at a time then the number of possible sequences of moves rises to 512^{10} or about 10^{28}. If we place no restrictions at all on changes to the arrangement of pieces on the chessboard, then there are about 4×10^{182} possible sequences of states of pieces on the board. That's 4 followed by 182 zeroes—a truly astronomical number.

What is true of chess is also true of most sorts of changes. For example, at any instant in time an animal moving through a landscape has a location, a speed and a direction. These values define its state at any one time. Its movements define transitions from one state to another.

The ball and stick model also applies to computers. The state of a computer is the net value of all the values in its registers. The program running on the computer decides what new state the current state will link to. So we can describe the behaviour of a computer program in terms of the ball and stick model.

This observation is important because it means that any process that we can model using a computer also has a ball and stick behaviour. It is beginning to look as if we can describe almost anything at all using the ball and stick model. This indeed turns out to be the case.

CASCADES AND CATACLYSMS

We have now seen that the ball and stick model can represent just about any system at all. It applies both to organisation and to behaviour; to forests and to fires; to chains of coral reefs and to nuclear explosions. It is universal.

So what? How does it help us to know that a coral

Figure 6.1 The connectivity avalanche in a random graph formed by joining balls with sticks. The diagrams show different stages at which the graph becomes saturated with clumps of increasing size. In each case notice that adding further sticks to join balls within different clumps immediately produces a clump twice the size, so clump sizes increase by doubling: 2, 4, 8 and 16 balls respectively. As the clumps increase in size their number decreases, so it takes only half as long to saturate the graph with larger clumps. This acceleration effect leads to the connectivity avalanche described in the text.

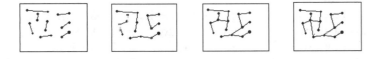

reef is related in an abstract way to a set of balls and sticks? Why should it matter that we can represent the behaviour of a computer in the same way? It is a truism in science that theories of everything explain nothing. On the other hand, it is also true that new ways of looking at things, especially if they relate apparently different things together, often lead to deep insights about nature. And this is the case with the ball and stick model. True, its descriptive power is limited, but at the same time it strips away all the messy details that confuse the issue when we look at different phenomena. It allows us to see the forest for the trees. It shows that beneath all that goes to make a coral reef different from a computer program is a deeper, simpler truth, one that is the same for both systems.

What the ball and stick model achieves is to reveal that what scientists have always studied as separate and

111

unrelated phenomena have a common underlying cause. It allows us to see the real nature of critical processes such as nuclear explosions, epidemics and the onset of chaos. In the early 1960s, two mathematicians—Erdös and Renyi—were studying the properties of large graphs. Now, a *graph* in this case is not a plot of *y* against *x*, such as a chart of stockmarket fluctuations. It is simply another name for the ball and stick model. What Erdös and Renyi found was that randomly constructed graphs inevitably contained what they termed a 'unique giant component' when the density of edges (sticks) exceeded a certain level.

To understand what this means, suppose that we have a number of villages in a large forest. At first there are no roads, so transport is difficult. Gradually, however, roads are built between some pairs of villages. As time goes on, some individual villages acquire a second road link. So it becomes possible to walk by road from village A to village B, and then by road again from village B to village C. Eventually, the system of roads becomes so widespread that we can walk by road from one village to any other village at all. The important thing in this example is not that all the villages eventually become linked when there are enough roads. What matters is the way in which the connections spread.

Suppose that for some set of villages A, B, C, . . . we can walk via a series of roads from any one village to any other. That is, there are roads for each pair A-B, B-C, C-D and so on. We can call a set of villages that are joined this way a 'cluster'.

Assuming that roads are built at random, then *at the beginning* each new road will join two villages, neither of which has roads leading elsewhere. So the size of the largest cluster will be just two villages. For a long time no larger cluster will form because the density of roads is just not great enough. Eventually, though, the region will become saturated with roads. At that point it becomes almost inevitable that a new road will link two villages that are already linked by road. So the new cluster is likely to contain four villages. Continuing this process we see that the region begins to fill with clusters of four villages. Moreover, because the new clusters are twice the size, it takes only half as long for the region to become saturated.

At this point a cascade effect occurs. Each time the region becomes saturated with clusters of a certain size, the next stage forms clusters that are twice the size. And because these new clusters are twice the size, it takes only half as long for the region to become saturated again. When still larger clusters start forming, they are again twice the size of the previous ones—four times the size of the earlier ones—and again it takes only half as long to saturate the region. That is, the pace of cluster formation accelerates very quickly. So once larger clusters start forming, they expand rapidly. As the size of the clusters grows, so their number decreases, until one cluster absorbs all the others. This super-cluster appears when one cluster gains a size that includes over half of the villages. From this point onwards, no other cluster can achieve a size equal to this *unique giant component* (UGC), as Erdös and Renyi called it. The process then becomes a matter of absorbing all the remaining clusters into the UGC. The rate of growth of the UGC falls off quickly until it has absorbed all of the villages.

What is true for villages in a forest is also true for any graph. Any time that we randomly add edges (or sticks) to a given set of nodes (or balls), there is a sudden phase change within the graph as a UCG forms. The nodes switch from being mainly disconnected to fully connected. Moreover, the process has some important properties. First of all, the number of edges that we have to add before a UCG forms is predictable. If we have N nodes then we need to insert close to N/2 edges on average before the UCG appears. Secondly, as the number of edges increases, the number of clusters at first increases until saturation point is reached. Then the number declines again as larger clusters absorb smaller ones into a single cluster—the UCG.

Thirdly, the composition of the system becomes inherently unpredictable as the UCG forms. To understand what this means, suppose that we perform an experiment where we fill in a certain number of edges in our graph. Then if we repeat this experiment many, many times we can say that a result is predictable if it comes up most of the time. If the number of edges that we add in the experiment is small, then the state of the graph is highly predictable: the maximum cluster size is nearly always two and most nodes are isolated. Conversely, if we add lots of nodes, then almost all the nodes will belong to the UCG. However, if we add around N/2 edges, the critical number at which the UCG appears, then both the size and the composition of the UCG are extremely variable.

Scientists measure variability using a number called the *standard deviation*. In this case we can use the standard deviation of the size of the largest cluster in the graph as a measure of variability. If we plot this variability against the number of nodes added, we find that its value is nearly zero when the number of edges is either very low or very high. On the other hand, it shoots up dramatically at the phase change where the UCG appears.

Finally, the traversal time for the system is maximum at the phase change. To see what this means, return for a moment to the example of villages in a forest. We can ask what is the longest time it takes to walk from one village to another. Let us refer to this time as the *traversal time*. A simplistic measure of this time is the number of roads we have to follow between villages. At first, when roads are scattered and never link up with others, the traversal time is 1. As the size of the clusters increases the traversal time goes up. When all the villages are finally linked into one giant cluster, the traversal time is large because many detours are necessary. The connections are not as efficient as possible. However, as still more roads are built, shortcuts become possible so the traversal time decreases. Now, taken in isolation, the above findings about connections in graphs are simply a curiosity. But the name *graph* is simply another name for the ball and stick model that we looked at above.

Remember that the ball and stick model is implicit in virtually any complex system. So the phase changes in connectivity that occur in graphs can occur in any complex system. Just what this means depends on what aspect of the real world the ball and stick model represents. To start with, recall the starfish outbreaks we discussed earlier. The ball and stick model is relevant because the balls can represent coral reefs and the sticks can represent the spread of starfish larvae from one reef to another. In this case the reefs are relatively well connected. The series of outbreaks could be interrupted if currents failed to carry the larvae from reef to reef. This might happen if the number of outbreaks were reduced or if the size of an outbreak at some key point was severely dampened. The same sort of property is true of epidemics. The phase changes in graphs reveal the wisdom of isolation—the traditional method of halting epidemics.

CHAPTER 7

PATTERNS IN THE SAND

The Mandelbrot set is one of the most frequently displayed mathematical images. It has appeared in many places, in art, in illustration and in mathematics itself. It is a *fractal*, probably the first to be identified as such to the public at large, and was brought to the fore in Benoit Mandelbrot's famous book, *The Fractal Geometry of Nature*. But the set named after him—the example of a fractal that everybody has heard of—doesn't look like anything from the natural world. Yet Mandelbrot, at IBM, and subsequently with colleagues such as Richard Voss, led the way to our seeing fractals as an excellent representation of many physical and biological patterns; in fact, they are almost the rule rather than the exception. People have been looking at fractals for geological time, so let's look at natural fractals.

UNIVERSALITY OF FRACTALS: EXAMPLES FROM THE REAL WORLD

Bob Mossel, a leading landscape photographer, has for many years been taking photographs of patterns in nature. As a keen amateur pilot, his photographs from the air reveal features and structures not obvious from the ground.

River deltas, sand dunes, salt lakes and many other

116

Figure 7.1 The Mandelbrot set. The set is actually a map of the border (phase change) between two kinds of behaviour in the formula $X^1 = aX = +b$, for complex numbers. The set contains many regions that resemble the entire set itself.

natural patterns are all examples of, on the one hand, self-organisation and, on the other hand, fractals. The process of fractals is endemic, a fundamental part of nature—in plants, sounds, animal markings. In fact they are far more prevalent in nature than the simple rectilinear constructs of man-made objects. Fractals appear everywhere because they are natural growth processes. They are characterised by scaling, endless repetition and growth of simple units, as shown by the example of the Blackwort fern popularised by Michael Barnsley in his book, *Fractals Everywhere*. As well as characterising static

117

patterns, fractals also characterise complex dynamics and the shape of chaotic attractors, which we will discuss further.

THE IDEA OF SELF-SIMILARITY

The basic idea is that of self-similarity. To understand it, let's look at the generation of the branch of a tree. Each branch develops buds which themselves produce new branches—each new branch a tiny replica of the old one. It doesn't have to be exactly the same. This process is captured in the L-systems we saw in Chapter 5. We can add some randomness to get something which looks more like real life, too.

There is one important difference, though, between any fractal model of nature and the actual phenomenon. There is only a finite range of scales for which the approximation is valid in the real world. Real objects don't keep getting smaller and smaller, as does the mathematical entity. Nevertheless, some fractals such as Brownian motion, do scale by factors of over a million. This property of endless repetition down to infinitesimal scales creates the special space-filling properties of fractals. They are almost like solids, but not quite—more like very fine sponges—giving rise to one of their special, well-known properties, that of fractal dimension. These are tricky mathematical concepts. Intuitively, there are many analogies in the natural world. Foam is a good one: it can occupy a huge volume but will condense down to just a tiny amount of liquid. Talcum powder is another: a fine dust which will cover a huge surface, and itself having a huge surface area.

Our childhood introduction to dimension is a very simple one. There are three spatial dimensions, with points, lines and solids having one, two and three dimensions. Later we might start to think of time (a

Figure 7.2 The famous Sierpinski Triangle. Notice the fractal property that the entire pattern is repeated in each subtriangle.

closely related quantity) as a further dimension. Fractals take us into the new territory of *non-integer dimensionality*.

The oft-quoted example is the length of a coastline. Suppose we were to start in Ullapool at the far north of Scotland and travel by small boat down the west coast of Britain to Penzance in Cornwall. We would, all being well, make it down there in a reasonable time; and if we had travelled at constant speed we could work out the length of the coastline by dividing the time we took by our speed. The coastline of Britain, however, is pretty ragged, particularly the west coast of Scotland. If we were to go into and around every loch, hugging the coastline very tightly, it would take us a lot, lot longer.

119

But our boat will be simply too big to follow the line of every tiny inlet. If we want to do that, we'll need a smaller boat. It should be easy to see, though, that as our boat gets smaller and smaller and the coast gets more and more jagged, we are never actually going to get to Penzance. So even though we have been travelling along a line, we're really attempting to cover an infinite distance.

Over the last hundred years, mathematicians have had a lot of fun with infinity. Not all infinities are the same size, according to their subtle arguments. The paradoxical features of fractals, infinite but not quite filling space, are examples of this hierarchy of the infinite.

Dynamic spatial patterns exhibit self-similarity too. Waves on the sea are such an example. Tour operators love to advertise small boats sailing in pictures of an island paradise. What could be more relaxing than lying out on the trapeze of a catamaran lazily sauntering around in the sunshine? The truth is often not so simple. If it's windy enough to lie out on the trapeze, there are also likely to be waves. Going upwind (at a slight angle, since you can't sail directly into the wind) causes the most trouble. As the waves hit, your small boat gets thrown around and you need to keep your wits about you to avoid falling off. Now, if all the waves are the same size a gentle rhythm is still possible, but they are not, and every so often a large wave will come along which will knock you off the boat. Waves come in different sizes. But although there seems to be a self-similar property here, mathematical models of wave systems are not necessarily fractal: thus some caution is necessary in assuming that every system with a range of scales is actually fractal.

Surfers also talk of big and small waves and trying to guess when the next massive wave will hit. Imagine now that a storm comes through and your island paradise becomes somewhat less idyllic. The wind churns up the sea and, particularly around the land, the waves end up coming from all directions. Now you are buffeted this way and that, just like a dust particle in the atmosphere being jolted this way and that by air molecules. Such a simple system hides a very interesting phenomenon, Brownian motion. The trouble with having a name like Brown is that, if you become really famous, there are other famous Browns to contend with. But as a scientist, you couldn't be much prouder than to be associated with the phenomenon of Brownian motion.

In 1827 Robert Brown observed this jitterbug behaviour in pollen particles in the liquid on a microscope slide. It arises through the random blows of air or liquid molecules exchanging energy with the particle so that it itself has the same average energy as the gas molecules. But the jittering looks the same at different spatial and temporal scales. Brownian motion is a standard part of statistical mechanics in most physics courses from school onward. But it has recently become of great interest because of this very property of self-similarity across different scales: Brownian motion is a natural fractal. If we plot positions of a Brownian motion particle as a function of time and space, the traces all look basically the same. But, unlike the pure fractals of the Mandelbrot set, there are wide variations and the traces are not exactly self-similar. The similarity is more *stochastic*: the traces look the same on average, this property being referred to as self-affinity.

BIG WAVES AND LITTLE WAVES

To go a little further into characterising Brownian motion, we need to understand the notion of the power spectrum of a random process. First we need the idea

Figure 7.3 Fourier analysis of square waves

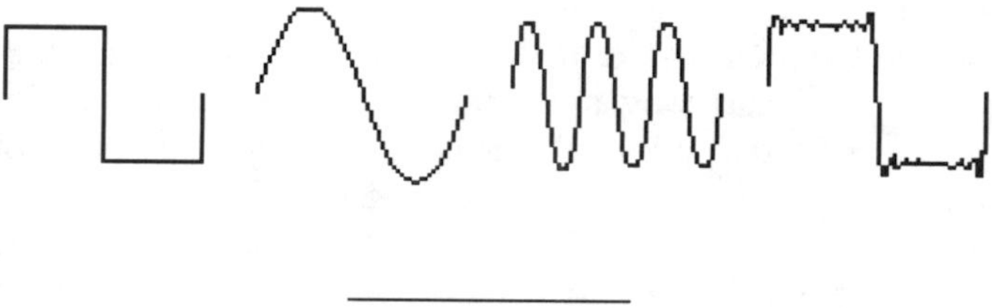

of *Fourier analysis*, a numerical procedure for the break-ing up of a signal into components.

The idea of waves and frequency is commonplace. We're familiar with waves on the sea, sound waves or even wave-like patterns in the sand. Waves are periodic phenomena, in which the up and down motion, or amplitude, varies in time or space. Just as a big sea will have a very deep swell or very large waves, the energy of a wave depends on the amplitude—the bigger it is the larger the energy. In fact the energy varies as the square of the amplitude, making large waves proportion-ately more energetic than smaller ones.

The brilliant insight of the French mathematician, Fourier, was that the mathematics of waves has much wider applicability than those systems that have an obvious periodic character. Leaving aside a few mathe-matical niceties, it is possible to build up almost any shape out of a series of simple waves, even if the shape has sharp edges. This seems counter-intuitive, but exam-ination of Figure 7.3 shows how this happens for the example of a square wave. To get the sharp edges, though, we do need a rather large (in fact infinite) number of components. The simple waves we use here have a particular mathematical form: as shown in Figure 7.3, they are trigonometric functions and are often referred to as sinusoids. A wave retains its shape as we change the frequency, scaling in just the way that a fractal scales. So it isn't surprising that fractals have a simple form in the Fourier domain.

With random or quasi-random processes, it is convenient to deal not with the Fourier analysis but with a closely related concept, the power spectrum. This tells us how much energy there is in each frequency component and is closely related to both Fourier analysis and the *correlation* in the process itself.

Linear systems, which we come to shortly, have special properties which make the wave analysis extremely useful. For example, the blurring of an image is simply a reduction in amplitude of the sine waves of very short period (high frequency) that contribute the fine detail to the image.

Fourier analysis gives us a neat recipe for making natural-looking fractals: the wave amplitudes all seem to decrease as the wave frequency gets greater in a similar way. All we need is that the Fourier spectrum fit within a particular range of shapes. Pure Brownian motion has a frequency spectrum that falls off as the square of the frequency, $1/f^2$, but models of Brownian motion often extend the frequency exponent to an *envelope* of values on either side.

Many natural phenomena have a physical structure in which the amplitude of the pattern falls off as frequency increases. That is, that size is related to $1/f$, the inverse of the frequency f. If we increase the exponent—that is, $1/f^3$, $1/f^4$ and so on—then the amplitude falls off even faster as the frequency increases. Thinking back to the building up of a square wave, we can see that the high frequencies, which change very rapidly, are what produce the sharp edges. So it is with different fractals. If the exponent is too high, then the results look too smooth to be natural. At the other extreme, they look too sharp and spiky.

We shall see later that there are models that underlie fractal patterns which helps to explain why they are so ubiquitous. But there is one very prominent fractal character that has proved much more difficult to explain. These are fractals whose power spectrums fall off as $1/f$. This is at the low frequency end of the range we saw for fractal Brownian motion, meaning that the patterns have more high frequency energy and hence seem sharper.

Examples abound. In physical processes they range from noise in semi-conductors to water levels of the Nile, from traffic on freeways to music. Across cultures and centuries, musical melody has this spectrum. The latest in musak is a fractal, synthesised on the fly, while fractal-derived music also forms part of serious artistic endeavours.

Common mechanisms behind natural fractals

Observing fractals in many natural systems leads one to suspect that similar underlying processes cause them. Remember that a fractal pattern is a global phenomenon. The aerial image of a particular river delta, for example, with its fractal branching pattern, is not the result of an observer tailoring this image. At the local level, as the river carves out its path, there is no knowledge of the big picture. Certain sorts of growth process simply generate interesting global patterns. One such process is *diffusion-limited aggregation*, or DLA, a ubiquitous and much studied process.

To understand DLA, we start with a growth centre in an environment or medium. Particles diffusing past are captured if they come within some given distance. They enter one by one at the outer circle, some distance

Figure 7.4 Examples of diffusion limited aggregation (DLA). On the left is an example of dendritic growth. On the right are structures formed by aggregation of falling particles subject to different levels of Brownian motion and different degrees of 'stickiness'.

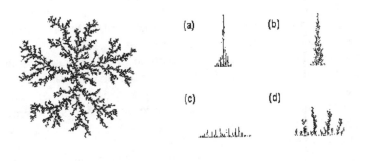

from the growth centre. They then move at random along the grid, going in any direction. Sometimes they go back outside the circle and are carried away. Others hit the growth centre, now a growing aggregation, and stick. The resulting dendritic growth has a fractal dimension of around 1.7, similar to phenomena in physics, biology and even the development of cities.

To explain the precise structure of the growth region in qualitative terms is difficult, but we can get some insight as to why a somewhat irregular mass develops. Imagine that we shoot particles towards the growth region from any direction, but in a straight line. If their starting positions are random, we will just get a disc growing out with rough edges. But now, as a particle staggers from side to side, instead of moving in a straight line, it sweeps out a cone and is likely to hit that part of the growing cluster sticking out furthest that lies within this cone. Hence the bit that is furthest out will grow faster

than the remainder. But because of the inherent randomness of the particle's 'walk', growth occurs elsewhere too.
This intuitive approach shows us that the randomness
in the walk (size of the cone) will affect the compactness
of the structure, and that if the particles do not stick
but have only some probability of sticking, different
forms will occur. Thus within this simple framework a
large range of patterns can arise.

Although we think of man-made phenomena as being
composed of straight lines and regular shapes, seemingly
random interactions between people produce other patterns. Studies of cities have shown them to have some
fractal shape or morphology. It's easy to see this as DLA:
the city grows as people drift in and settle. They are
most likely to settle where there is already some habita-
tion. This of course ignores the expansion of the city
within any specific controls exercised over development.
Nevertheless, the results fit many older cities, as
described by researchers Michael Batty and Paul Longley
in their book *Fractal Cities*.

Similar mathematics describes the very opposite process to DLA. It is usually referred to as *dielectric
breakdown*. The branching structures we have noted
might remind you of lightning. Rightly so, since the
forked structure of lightning strikes *is* fractal in character.
Great charges of static electricity build up on clouds—to
the point where the resistance of the air breaks down, a
huge discharge occurs and the charge flows to ground.
But once the breakdown has started, many new breakdown points arise along the way, giving rise to the forks.
The material (air) and the nature of the field (clouds
etc.) govern just how the discharge takes place, and again
we can get a wide range of patterns from this simple
process. So here we have the opposite model for the
development of a city: expansion pressure from a single

point. Still a fractal phenomenon, but not a cause of DLA.

Thus the goals of complexity research are achieved: models of physical processes, the underlying mathematics, extending to patterns of social development!

A new tool for engineers and scientists

Dot puzzles are very popular with children—join the dots to make a well-known shape—but deciding which dot to join to which involves powerful pattern recognition mechanisms. If the dots are not evenly spaced or are crowded together, the interpolation of the contour in the gaps may be quite difficult. Our sensory systems are always looking for patterns—the pictures in the fire and ink blots mentioned above, for example. Our mistaken perception in visual illusions of all kinds arises because we try to make sense of image data, even when it's somewhat unusual. With fractals so pervasive in the natural world, one might expect that we make use of them for pattern recognition. Although it seems that we are fairly good at recognising spatial transformations of objects, we are rather limited to common movements of solid objects in 3D space. When we classify a pattern we note the elements, but how precise is our recognition of the interrelationships? Would we notice deviations from self-similarity in a fractal structure? At present it seems that we would not, which is quite surprising.

Interpolation is the filling in of the gaps between data or sample points. Traditionally, the methods interpolation have emphasised smoothness, making sure that nothing too wild happens, an essentially conservative approach—and that is what we would normally do in our child's puzzle. But fractals are far from smooth. So if we know or have strong reason to believe that a set of sample points is drawn from a fractal, then the interpolation function should be quite different and should not be smooth. This is a novel area, with much work to be done in it.

Another way to think of this is in terms of the shortest descriptive computer program needed, along the lines of Kolmogorov complexity discussed in Chapter 2. If we have a ragged set of data points, defining a contour through them will be very complex with many line segments to list. But if we know that the raggedness is part of an underlying fractal process, we can get by with a much simpler description. This same approach leads us to think of new and much more powerful ways of compressing data. If we have objects of low Kolmogorov complexity, it makes more sense to recreate them from scratch rather than to send copies of the images.

A fractal is a complex structure, but it is often made from very simple generating rules. Thus instead of sending somebody a copy of a fractal image it takes much less bandwidth or storage to send the generating program. On the other hand, we do not get something for nothing. We have to put a considerable amount of work into designing the program in the first place. Michael Barnsley specialises in the fitting of fractals to images. His techniques, described in his excellent book *Fractals Everywhere*, apply even to pictures one would not at first think were fractals. The method rests on the Collage theorem, which captures neatly the idea of self-similarity. We start with a fractal object: Barnsley's well-known example is a fern. We now look for a geometrical transformation that will shrink and maybe distort the original image to produce a tile. We now try to make a set of such tiles that cover the original picture: the more exact the coverage, the more exact the approximation. Now the set of transformations form an iterated function system, which we can use to regenerate the original image.

Back to the travelling salesman

In Chapter 2 we discussed the Travelling Salesman Problem, highlighting the difficulty of solving the problem for large numbers of cities. But much effort has gone into finding simple approximate methods. One interesting example is the use of a fractal, the Peano curve. The fractal dimension of this curve is 2; that is, it fills the entire circumscribed area. The approximation is as follows. Each city will lie on a particular point on the curve. Now sort the cities according to where they are and that will give an approximation to the shortest route. To see how this works intuitively without a mathematical proof is not easy. At any point the Peano curve grows outwards like a locust swarm devouring every point in its path. Points close together on the Peano curve are always close together in normal coordinate space. So as it sweeps outwards, it collects the cities more or less in increasing distance from the start. But if we were to order the cities according to concentric circles, we would end up with errors where two points close together along a radial line were visited in different sweeps: the wriggling path of the fractal makes sure they all get carried along at the right time. Does this theoretical method have any relevance to the practical world? There is no obvious way in which cities adapt, so it is hard to see any sort of hidden evolution of cities: we have only the specific planning experience of city architects. But it remains the case that a road or transportation network similar to the Peano curve will facilitate delivery systems by providing natural optimisation of travel times.

Since Benoit Mandelbrot wrote his famous book on fractals, we have seen their appearance throughout nature. From complicated shapes and patterns we now turn to complicated *dynamical* processes, but we shall see fractals turn up again there too.

Figure 7.5 Formation of Peano's space-filling curve. The first diagram shows the basic pattern. Subsequent iterations repeat the basic pattern at ever finer scales.

THE CHAOTIC ROLLER COASTER

In the 1980s, *chaos* became a fashion, a household world. Not the chaos of everyday life, though, but a property of dynamical systems. The strange patterns of motion, and the notion that systems could be unpredictable even if there were no noise, captured the public imagination. Chaos became a topic for conversation among sociologists just as much as among ecologists.

Before we can appreciate the nature of chaos, it helps to understand a little of how simple, linear systems behave. Two central ideas in linear dynamics are stability and equilibrium. Some simple experiments may help the reader to understand what these terms mean. First, have you ever tried to balance a pen on its end? The pen needs to have a flat end, otherwise the task is impossible. Anyway, once you have succeeded in balancing the pen on its end you have achieved an equilibrium. Once it is in position the pen will quite happily stay where it is. Now try nudging the pen near its tip, ever so slightly. The pen immediately falls over. The balancing pen is an example of an *unstable* equilibrium. Any shift away from the equilibrium position *increases*, rather than decreases.

Next, instead of balancing the pen on its end, try dangling it on the end of a piece of string. If you suspend it vertically then the pen stays quite happily where it is. Again, it is in equilibrium. Now try pulling the pen slightly to one side and let it go. Immediately it starts swinging back and forth. However, it eventually comes back to rest in the vertical position. This vertical, resting position is a *stable* equilibrium. Any disturbance is soon erased by the system.

Non-linearity often arises through interactions. For instance, a simple pendulum behaves in linear fashion and its motion is predictable. But a certain popular toy, in which a pair of pendulums are attached to opposite ends of a rotating arm, behaves in a highly irregular and unpredictable fashion. In short, it is chaotic.

Non-linear systems differ from linear systems in several important ways. One difference is that linear systems tend to obliterate minor variations whereas non-linear systems magnify them. Consider the following examples. You are taking a bath and splash the water. This action makes a lot of movement on the water surface and makes the water depth very uneven. But after just a minute or so the water surface settles down and becomes flat and even again. Suppose that the water becomes cold and you add hot water from the tap. Then, as you add the hot water, the water temperature becomes very uneven. However, even if you do not stir the water it will soon reach an even temperature everywhere.

Both of these examples are cases where physical processes tend to obliterate local differences. They drive the level and temperature of the bath water towards an equilibrium. When the bath is in equilibrium it stays there unless you disturb it, such as by splashing, or adding hot water. In contrast, imagine a huge gas cloud out in space. If the gas is dense enough then the cloud will not dissipate out into space. Unlike the bath water it does not settle into an equilibrium. Instead it shrinks, because gravity draws the molecules together. As it coalesces there are minor variations in density. Instead of becoming smoothed out these irregularities contain centres of attraction that draw the gas inwards. Eventually these centres become stars and planets, and give rise to all the variety of objects that we see around us.

Examples of processes that magnify minor differences abound in biology. In the eye, for instance, nerve cells not only fire messages when stimulated by light but sometimes inhibit the response of neighbouring cells. The result is the tuning of cells, say to edges in an image. During early foetal development and growth, neighbouring cells in the embryo can give rise to completely different organs.

In all of the examples the enhancement of variations emerges out of countless interactions between simple objects. These objects are gas molecules and dust particles in the first example, and cells in the second. Interactions give rise to non-linearity. In general, it is non-linear processes that tend to enhance irregularities. This point has been emphasised by the Nobel Prize winner Ilya Prigogine. He examined the properties of 'dissipative systems' that are far from equilibrium and stressed that the potential for minor irregularities to grow into large-scale patterns is an important general property of these systems.

Irregularities grow into large-scale patterns because non-linear systems are sensitive to initial conditions. This effect was discovered by Edward Lorenz, when he tried to simulate weather processes on an early computer which we describe in more detail in Chapter 7.

It is this sensitivity to initial conditions that makes chaotic systems inherently unpredictable. This is obvious when you think about it. The essential problem is that we can never measure any variable with complete accuracy. Nor can we calculate the value of any parameter with absolute precision. These small errors are enough to ensure that the true behaviour of a system will soon deviate from any model that we use to try to predict its behaviour. So long-term weather forecasting is just not possible. Nor is it possible to make accurate long-term forecasts of (say) stockmarket prices, or about any other chaotic system. Several biologists have speculated that chaos is an important source of novelty in living systems. The point is that minor, local differences can lead to unpredictable, large-scale changes from the status quo.

PHASE SPACE

To understand the interesting patterns produced by chaotic systems, we have to take a special perspective, called the phase-space portrait. In our sketches of Brownian motion, we plotted the position of the particle against time. We could, however, determine its position and momentum at every moment and plot these against one

133

another, with each point on the trajectory representing a particular time. This combination of velocities and positions at any point in time completely describes a dynamic system. It is this phase-space portrait where we can see visually the interesting things that happen in chaotic systems.

Just as we see the pattern created by a spider's web by looking at it across space as a temporal history, so it can be useful to view many complex systems as a pattern—but not just a pattern in real space. In an ideal gas we have many molecules travelling at different speeds. To characterise each molecule at any given time we need to know where it is and where and how fast it is going. It is convenient to build a representation using both these quantities—the position and velocity vectors. If we consider a rarefied gas with just one molecule then we can plot a trajectory in *phase space* as a function of time. This path will be a series of straight lines parallel to the position axes (because the velocity is not changing), with sharp turns as the velocity changes at a collision with a wall.

We can now add more and more molecules, producing a cloud which is a sort of hyper-cylinder along the position axes but with velocity concentrated in a particular band with fuzzy edges. But for many molecules it is useful to think of the phase space as composed of separate axes for each and every molecule. Thus for N molecules we have 6N axes. For a gram of hydrogen, N is of the order of 1023, making this rather a large graph. Each state of the gas, every set of positions and velocities, has its own point on the graph, so we can again plot a trajectory of the whole gas as time goes on. This is a very complicated path. It's a thick knot and if we wait long enough it will fill a huge volume of phase space, although it will take rather a long time to get out to all the velocity extremes.

This random trajectory is actually not that interesting. But in some dynamical systems this trajectory develops some special structure, enabling us to describe it in a simpler fashion. It is these systems that are of interest to us in this book. In some cases the simplification is dramatic, resulting in a military-like ordering of all the elements of the system. In others, the trajectory still looks random, but has some special mathematical properties.

This phase space view is valuable for understanding chaos and many aspects of complexity. But for some of the biological systems we will consider it's somewhat limiting because the number of agents (particles in the gas, for example) is continually changing. This is the frontier: observable, but so far intractable.

Attractors and patterns

Dynamic systems have three types of attractor:

1   *A fixed point* The system reaches a point in phase space and does not progress any further. It remains at the same point in phase space until the end of time.
2   *A periodic orbit* The system enters an orbit, visiting the same set of points in phase space over and over again.
3   *A strange or chaotic attractor* Here the system, like the Flying Dutchman, is compelled to wander forever in phase space, never returning to the same point. The patterns traced out in phase space are visually intriguing and have stimulated artists as much as scientists. However, the attractor is very definitely different from a completely random system, which would visit every point in phase space. It is restricted to a limited volume. Furthermore, this volume has limited dimensionality. The attractor is a fractal.

135

Figure 7.6 Attractors for logistic growth

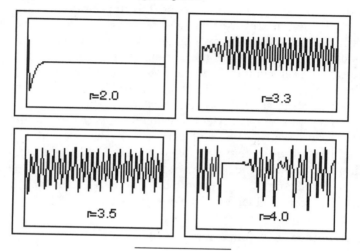

Free will returns!

Why is chaos so interesting?

Everybody, especially the English, know that weather is chaotic. What, then, was so special about Lorenz's discovery of chaos in weather systems? Like many discoveries, it lay fallow for some years, but chaos was to sweep way outside obscure academic journals to become the source of many popular books. Not only was chaos totally different, not only did it generate new and interesting graphics, but it also struck at a much deeper level. Chaos attacked the very heart of determinism and opened up new opportunities for free will!

Before the chaos boom we imagined that, if a system was free of random elements, then, if we knew the starting conditions, we could predict its evolution for ever more. That is still essentially true, but we now know that the specification of the initial conditions has to be impossibly accurate in the case of chaotic systems.

Figure 7.7 The Lorenz attractor. The lines show changes in a model weather system through time. This strange attractor is actually a three-dimensional pattern. Notice that the lines never duplicate the same path twice.

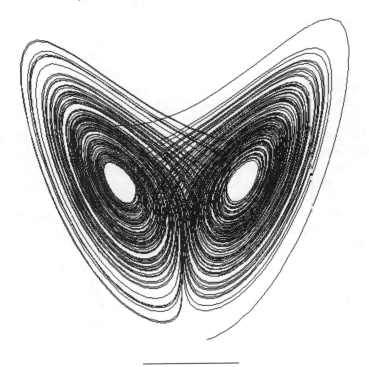

The weather

Although the famous nineteenth-century mathematician, Poincaré, had a deep understanding of many of the non-linear phenomena that make up chaos theory, the start of the initial wave of interest is usually considered to be the meteorologist, Edward Lorenz, at MIT. It isn't really surprising that the weather should be hard to

137

predict: land surfaces are irregular in shape, mountain ranges are far from easy geometric shapes, and water surfaces vary in size, shape and position without obvious pattern. But it isn't this cluttered nature of the world which interested Lorenz. Weather models in today's meteorological offices have many variables, easily as many as a million: Lorenz's model had just three.

The weather patterns on the earth have two main driving influences. Air is heated at the equator and moves out to the poles, from where cold air returns, and the rotation of the earth creates prevailing westerly winds. The earth's rotation has another effect, through the so-called Coriolis force,[12] creating eddies in the atmosphere as the hot air flows into the higher latitudes. It is this large-scale movement of air around the planet that Lorenz, like all good physicists, tried to capture in the simplest possible model. His phase space had just three variables and contained chaotic attractors: orbits in phase space which would travel on and on without rest. Two attractors that started close together would diverge rapidly.

The curious shape of the Lorenz attractor is reminiscent of a butterfly and perhaps for this reason the sensitivity to initial conditions came to be known as the 'butterfly effect'. But there was also another idea lying underneath the surface. The extreme sensitivity to initial conditions of chaotic systems means that the very slightest change in those conditions can produce radically different results. Such a tiny change might be evoked by a butterfly landing—an infinitesimal effect.

THE SCIENTIFIC BLINDSPOT

Although science proceeds by rigorous testing of theory, sometimes its progress slows down for one reason or another. Stephen Jay Gould, in numerous highly readable books, documents the erratic course of science towards

138

better paradigms. Sometimes the blockages are social (such as political pressure which propped up Lamarckian evolutionary theory in Russia), sometimes technical.

In this chapter we've seen how technical advances in computers have brought about breakthroughs in how we look at dynamics and how we describe the objects of the real world. By not trying to make things linear, so that one can do the calculations, nor trying to find simple smooth functions to describe shapes, we have found untold richness of dynamics and form.

In the nineteenth century, pure mathematician Georg Cantor became interested in sets that are like fine dusts, the predecessor to fractal theory. Early in the twentieth century, mathematicians studied sets from mathematical curiosity that we now recognise as fractals. These pure abstract structures underlie patterns and dynamics of the natural world.

Science tends to be a little bit like the drunk who is looking for his wallet under a street lamp at night. When a stranger comes up to help and asks where he lost the wallet the drunk explains that he lost it further down the street. He's searching under the street light because it's the only place where he can see.

The point of this story here is that a lot of scientific research is driven, or limited, by the tools available. From Galileo turning a telescope on the heavens, to DNA recombination, new analytic methods in science have always led to new discoveries. One reason why science did not deal with complexity long ago was that an essential research tool—the electronic computer—was not available. Even today there are some problems that the fastest supercomputers cannot tackle adequately.

CHAPTER 8

IMPLICATIONS FOR COMPUTERS

In the early 1990s, in the border zone of Italy and Austria, a corpse was discovered, not a climber of our time but a fully clothed traveller from the Iron Age, who died thousands of years ago. Imagine the excitement of anthropologists at such a discovery—tangible evidence of what life was like at that time. What would be the equivalent momentous discovery in computing? The idea of machines that think is now somewhat old: machines do think in limited domains; they play chess better than most of us; they can find correlations in huge databases beyond the grasp of any human mind. But could they ever be conscious? Will we ever see HAL, the computer of the movie *2001*?

Despite the advances in our understanding of neural networks, many people still believe that consciousness will never be a property of non-biological systems. Philosophical argument continues to rage, but as technology improves we can ask more and more precise questions. So let's take a look now at biological computation, or biocomputing.

THE HOLISTIC APPROACH TO THE BRAIN

Animal brains don't look a lot like computers. But are the underlying computational principles the same? This profound question has occupied neuroscientists and philosophers for decades. At the close of the millennium, controversy still rages on whether we should look at the brain nerve cell by nerve cell or as a single large adaptive system. Somewhat less influential at present are those, like Roger Penrose whom we've already mentioned, who believe that classical mechanics is not sufficient to model an animal brain.

The human brain contains approximately 10^{10} nerve cells, or neurons, each with up to 10^5 connections to other neurons. It's a giant parallel computer. For something like 30 years, the model of neural computation was a simple one: the signals coming into a neuron from other neurons were added together; if they exceeded some threshold, the neuron would fire a voltage spike to some other set of neurons (often including some of the input neurons). So the number of spikes fired per second would depend on the average number of spikes coming in. Huge networks take in data from the senses and ultimately fire spikes to muscles controlling movement of the body. In the human brain, of course, a lot of activity can circulate within before any muscular output appears.

Emergent properties of neural networks have come to the fore in the last few years but as an idea it's quite old. In 1977 Walter Freeman published a book entitled *Mass Action in the Nervous System* in which he argued that important properties of the brain are captured in the oscillatory behaviour of populations of cells. In particular, sets of connected oscillators will phase-lock under the right conditions, as we saw in Chapter 3.

This view of the operation of a brain was and still is somewhat radical. The conventional wisdom held that the activity of a nerve cell increased as its input increased. The activity is measured in terms of the firing rate of the cell, the number of voltage spikes fired per second. A vast amount of neurophysiology and theoretical modelling has

141

been devoted to understanding single cell activity. We now know in great detail the properties of cells that travel from the optic nerve to the brain; we know there are groups of cells that respond to oriented lines, colour, particular directions of movements, and even faces. It took rather longer for the scientific community to accept that these cells deep inside the part of the brain that handles vision were indeed sensitive to *faces* and not just oval shapes or other primitive figures. But a decade of experimental work has gradually turned the critics around.

The activity of a single cell is a discrete, stochastic signal. But, looking at a population of cells, the average will be much smoother, and with appropriate connections between populations the average activity will oscillate. Collective activity appears as phase-locking of these oscillations. Freeman saw phase-locking as a *major* element in the activity of a brain. His theoretical ideas were derived from experimental work on the olfactory bulb of the rabbit.

A more recent discovery, made by a former student of Freeman's, Charlie Gray, attracted a lot of interest in this rather quiet area. Gray and others looked at the firing rates of cells in the areas of the brain where visual information is first processed, the striate cortex. There are numerous different types of cell in this area. The kinds Gray investigated each look out into their own particular region of space and respond when a line at a particular orientation appears in that region. Looking at the behaviour of just a single cell shows a noisy signal varying roughly linearly with the contrast of a line. But oscillations in the firing rate also occurred when long lines were presented to the visual field, exciting numerous such cells. Even more interesting, these oscillations in the firing rate were found to be phase-locked. This phase-locking could even extend to cells in either hemisphere of the brain. This discovery immediately suggests that the phase-locking, as a collective property, is the way in which the brain encodes an extended line or border.

This result was potentially big news: it is easy enough to record from a cell and deduce that it signals some simple pattern like an edge or a patch of red. But how are more complex signals represented? It's easy to see that the number of ways we can combine primitive tokens together must get far bigger than even the vast number of cells of a biological brain. The information we store is far bigger than the number of cells, just as the DNA of an animal does not in any sense describe the grown animal. So here was an exciting alternative: these complex structures are not represented by single cells, say a long twisty line cell, but by a group of cells along the twisty line which are all operating in synchrony with one another—jumping up and down together in a long twisty line. At the time of writing these issues are still hotly debated.

The coupling of the brain across the two hemispheres is interesting in another way. The two halves of the brain are just the top of a complex hierarchy of modular structures. Throughout the book we have seen ways in which complexity is brought under control by building hierarchies, in computer programs and in life. The brain is far from a huge homogeneous network.

If we look at the human cortex we find it is made up of perhaps a hundred such modules or cortical areas. At this modular division computer science meets evolution. In software engineering and in the development of computer systems generally, breaking up tasks into separate independent modules with defined interfaces between them is the key to managing complexity. Yet the brain clearly derived its modular structure during evolution: new modules developed as brains became more

143

complex. Thus if the brain just evolved that way, it would be hard to say that it has a modular structure for reasons of efficiency. The real situation is undoubtedly a complex mixture of the two.

There is abundant evidence that the brain's modular structure is genetically laid down. As the infant's brain develops there are critical periods in which some structures will grow or not grow depending on the environment. There is simply not enough DNA to 'order' the complete layout of a brain, but there is some complex mixture of direction of growth and experience laid down.

A common everyday experience of this is the learning of a second language. Beyond a certain age learning a correct foreign accent just seems to be impossible. In his highly informative book, *The Language Instinct,* Stephen Pinker points out that Henry Kissinger, former US Secretary of State and a man of undoubted intelligence, retained a strong accent, whereas his slightly younger brother speaks American English flawlessly. Joseph Conrad, a major novelist writing in English, learned the language at the age of twenty-five and, despite his literary genius, retained a heavy accent throughout life.

Like many profound ideas, the idea of innate knowledge is not new. Noam Chomsky caused a furore in the 1960s by propounding just such a suggestion:

> If the conclusions of this research are anywhere near correct, then humans must be endowed with a very rich and explicit set of mental attributes that determine a specific form of language on the basis of very slight and rather degenerate data.

In fact Stephen Pinker describes striking empirical evidence some three decades later, with the discovery of a family with a defective language gene. Its carriers are unable to develop some simple grammatical constructs.

144

Exciting though this is, the numbers of individuals encountered is small, suggesting that we should be cautious in relying too heavily on these data.

Thirty years later we know a great deal more about neural systems and learning theory, but natural language processing is still some way in the future. These questions of innate knowledge and consciousness are the very frontier of neuroscience.

We discussed modularity and object-oriented computation in Chapter 2: in a brain we see the ultimate hierarchy of computational modules. So, having looked briefly at the relatively recent holistic discoveries, let's now dig a little deeper into the nature and action of neurons themselves.

BIOCOMPUTING

In the 1960s artificial intelligence was strongly oriented to rule systems and inference methods. There were some notable successes in (say) expert systems, but in other areas such as pattern recognition and robotics progress was much slower. It always seemed unsatisfactory that things a tiny insect could do should defeat the top supercomputers. It was frustrating to engineers to find that sharks have evolved a *rough* skin surface to *reduce* drag in the water. Thus it was not surprising that eventually people should turn to studying how animals and ecosystems 'compute'. In the late 1980s the trickle of engineers and scientists moving away from rule-based systems, so-called strong AI (artificial intelligence), and into biocomputing became a torrent.

Neural computation

A late twentieth-century computer chip will have something of the order of ten million transistors. Each

145

transistor has just a few input connections. A human brain has 100 to 1000 times more components, each with 10 000 connections, and with much more sophisticated processing of all the inputs. Yet universal computation as defined by Church and by Turing is possible with just simple digital circuits. There are still plenty of people who believe that digital computation cannot emulate the computation of a human brain. This may be true: but we still do not know anywhere near enough about the computation of multi-agent systems to be sure that mental phenomena do not fit into this classical picture. But despite the immense sophistication of today's computers, we have a simple model, the Turing Machine, of computation. We do not as yet have such a definitive model for neural networks, although there are many suggestions.

Hodgkin and Huxley received the Nobel Prize for elucidating the basic principles of a nerve cell, or neuron. It is a very sophisticated system involving several different ions (principally sodium, potassium and calcium) travelling through gates in cell membranes. The gates can be opened and closed while ion pumps in the cell membranes will move ions across, against a voltage gradient. More recently, Penrose has brought increased attention to bear on the detailed, so-called microtubule structure of a membrane itself and the computation (possibly very significant) going on there.

Modelling nerve cells at such a level of detail does not allow one to study large networks, not with present computing resources. So this lowest level of computation is approximated in many biological studies and almost completely ignored in artificial neural networks.

It is difficult to know where to start in building a model neuron, since almost every simplification we might make could lead us into trouble. So, very broadly speaking, a neuron receives inputs from many other neurons at the synapses of its dendritic tree. These inputs are summed, more or less linearly, to produce a single voltage output. One of the first modelling assumptions is that the output should be either zero or one, a Boolean value. In fact real neurons are of two types. There are those with a continuous output, which is essentially a voltage just like a signal in a man-made electrical system. But most of the neurons that make up a mammalian brain are spiking neurons. They work by generating voltage spikes or pulses. These pulses are all roughly the same size, and the rate at which they occur in the spike train indicates the strength of the neuronal output. This is much closer to the Boolean output, but the signal being conveyed is the fraction of time the unit is *on* (spike) rather than *off* (no spike), as opposed to a yes–no vote. This brings in the complication of time dependency.

A real neuron is a physical system evolving continuously in time. Simple neural models frequently are discrete, similar to a cellular automaton where the updates to each element occur synchronously in a stepwise fashion. On the other hand the voltage spikes can occur at any time, not just at specified time intervals. We don't yet know the full importance of asynchronous versus synchronous processing. But we do know that it makes a difference. Cellular automata do not produce the same patterns if the updates are done randomly as opposed to all together. A recent study by May and his colleagues showed that evolutionary systems also look different in these two situations. Even more exciting is large-scale temporal order, which we consider below. Another simplification is to assume that the computation of output from the input is done to high accuracy. Nothing could be further from the truth: real neurons are 'noisy'. Signals are not always totally reliable and may vary slightly in a random way. One might think at first that taking out the noise in a model system couldn't possibly be wrong—the model would just be more accurate. But remember the discussion of Synergetics in Chapter 3. The fluctuations are essential for order to arise.

Our abstract model of a neuron usually has no spatial distribution: it doesn't need one if all that is required is the computation of an output signal based on a set of inputs. Real brains are a spaghetti-like mess of inter-twined cells, essential simply to create the right connections. Now signals which come in from other neurons a long way from the centre of the cell are weakened somewhat on the way in. Furthermore, inter-fering signals (so-called inhibition) can come in on other branches of the cell. The branching patterns are fre-quently fractal and although this could arise just as the result of the growth process it might have some impor-tance in the connection strategy. But we don't know yet if this is fundamental or just an inevitable feature of the biological constraints. Walter Freeman puts this suc-cinctly: a human brain has to live for close to a hundred years and a lot of its construction is concerned with biological maintenance over that time scale.

Learning versus innate knowledge

We have made tremendous progress in the last decade in understanding how learning occurs in neural systems. But artificial systems still differ fundamentally. Mostly they are supervised with error feedback immediately a mistake is made. Some systems do use reinforcement learning, in which feedback is given only occasionally and then in an indirect way. A good example of this is Tesauro's backgammon program. It has to make many moves before it gets just the one piece of data fed back to it: that it has won or lost.

In biological systems, learning takes place at the

synaptic junctions between neurons: precisely how it occurs is still a subject of research, although some proposals such as Hebb's of 1949 still account for quite a lot of neural learning. In the Hebbian model, synapses are strengthened when the neurons they connect fire simultaneously. Thus the neural system builds up a sensitivity to repeated coincidences. But in itself this scheme is inadequate to account for all that goes on: it does not allow for weakening of synapses and it ignores the several different timescales over which learning occurs.

The brain is also full of chemicals, small protein fragments which influence neural communication. Some directly control synaptic activity. Others are more obscure, leading to changes on a more global scale in strange ways. Even nitrous oxide, a reactive, highly poisonous gas, plays numerous roles in the brain. For example, an excess of dopamine is a characteristic of schizophrenia, while a deficiency of it is linked to Parkinson's disease. But although the correlations are well established, neither the mechanism nor even 'which causes what' is certain at this stage. Again we ask whether this sophistication represents an accumulation of tricks during evolution or whether the system would not work, or not work as well if it were simpler.

Training a large homogeneous network to do a very complicated job is difficult. For some simple artificial networks it is very difficult indeed. Judd has shown, for example, that for a class of popular artificial networks it takes exponentially increasing time to get a network to perform bigger and bigger tasks, just as that Travelling Salesman Problem blows out of control as the number of cities increases. That is a real killer, and would indicate that we are never going to be able to build large networks of this kind while they are homogeneous. Thus the need for modular structure, piece-by-piece, task-by-task learning is essential from a computational perspective, mirroring the recent trends in object-oriented programming.

Before leaving the intricate details of neuroscience, we might take in a cautionary tale from a quite different area. Researchers at the University of Sussex have been evolving hardware. They start with a chip referred to as a programmable gate array, in which the connections between its many elements can be user-configured. They use evolutionary programming, which we come to shortly, to evolve the configuration of the chip. It works very well. But the really interesting thing at the time of writing is the effectiveness with which the evolved solutions make use of subtle electrical properties in the chip.

It would not be surprising if it takes us a very long time to unravel all the interactions and synergies in biological brains.

Since Noam Chomsky's famous conjectures on the innateness of language there has been the suspicion that we are born with some predispositions. Chomsky received a great deal of criticism of his proposal but it is now fairly widely accepted. But the human's genetic code is nowhere near large enough to code individual connections' strengths between nerve cells, so this innate knowledge must take some more subtle form. It could be in the large-scale organisation of the brain into different areas. This large-scale organisation does persist from individual to individual in a highly consistent way, so it does seem to be under genetic control to some extent, but local constraints are likely to be involved in some way.

One difficulty with the 'determination of areas' argument is that the cortex is highly malleable. The somatosensory cortex, the area which processes tactile information, exhibits a clear map of the body. But the map is distorted: some parts of the body, such as the hands and fingers, get much more cortex than their simple skin area would suggest. This is not surprising since we have much greater sensitivity and functionality in our fingertips than in the small of the back. Following an accident, say the loss of a finger, some cells will no longer be receiving input. But the cortex now reorganises very rapidly, and the area allocated to that finger rapidly gets taken over by the adjacent fingers. So even if this structure is laid down genetically, it can be very easily changed at any stage of life.

Blood clots within the brain, strokes, cause massive, localised destruction of neurons. But victims, at first badly incapacitated, often make a remarkable recovery, indicating the continuing plasticity of the brain late in life.

In an intriguing experiment, the optic nerve has been redirected during the early life of the subject to some other area of the brain than that which would become the visual cortex. Surprisingly, the visual cortex develops normally but in a different place! So we get the impression that the cortex of the brain is a general purpose processor: what it does depends upon its input and perhaps what is demanded of it. So, returning to Chomsky, perhaps what we have is restrictions on the computing power: only some sorts of operations are possible, and perhaps this is built in at a very low level.

We can get an inkling of what happens from a very interesting idea proposed by Jim Stone. He suggested that the synaptic learning rules are set up in such a way that they are tuned to temporal invariances, things which do not change very much over time. He demonstrated this successfully with a neural network which, without any supervision, learned to recognise stereo disparities. There is a lot still to explore in this exciting area.

Finally, we can ask how the learning rules of a biological neural network themselves evolved. It would seem that there is a driving force to find structure in

151

the world: we are always trying to make sense of the flood of data that comes in through our eyes and ears. But nerve cells are expensive to run. The brain consumes as much as 20 per cent of the body's energy while at rest. Taken together these observations suggest that the neural learning rules might have evolved to develop the simplest and smallest networks to do a job, rather in the manner of finding the algorithmic complexity of an object.

Since we can now make extremely tiny computer chips with awesome performance as number crunchers, data miners, machine controllers, pattern analysers, and many other applications in daily life, we might ask why neural computation evolved as opposed to the digital computers. Of course there is never any answer as to why something evolved. Evolution is contingent and highly uncertain. But Abu-Mostafa at Caltech has an interesting argument as to why neural networks are better than the high speed sequential processors we have sitting on our desks.

He introduces the notion of random problems. In such a problem the rules to solve it are *very* lengthy and complex. Consider distinguishing oaks from ash trees. The shape, size, colour are very irregular. Leaves on one particular tree are similar (to a human!) but that similarity is very difficult to capture mathematically. In such a problem the list of rules and exceptions can be more complicated than just listing all the examples. In other words, these are problems of high algorithmic complexity: they require long programs to solve them. Here, then, is Abu-Mostafa's conjecture: that neural networks are better set to solve problems of high algorithmic complexity than conventional computers are, although they lose out by a long way on speed.

As we learn more and more of the nature of neural computation we will be able to test this idea more thoroughly. It is attractive, though, to think that biological brains may be highly optimised for their role.

EVOLUTION AS A NEW TECHNIQUE IN COMPUTER PROGRAMMING

Just as neural networks have swept through the computing world, so evolution is poised to break into the world of applications after being a minority interest since the 1960s. The area in which it has found immediate interest is the area of computationally hard problems in optimisation (Chapter 2). The principle is simple: encode the problem solution in some way as a genetic code, then evolve the solutions according to some quality level which becomes the fitness of each phenotype. We know that evolution over time has produced some very effective solutions. But is it fast compared to more conventional methods? It is indeed fast up to a level of reasonable approximation, but not when a perfect solution is required.

There are numerous practical difficulties, however. Choosing the population size, methods of selecting the fittest individual and, most importantly, choice of representation all make the method far from guaranteed to work. Nevertheless, the range of successful applications is extensive: job scheduling, timetabling, cellphone frequency allocation and many more.

Finding an optimum

In the days of the first exploration of our planet, intrepid adventurers would hack their way through forests, ford rivers and brave whatever hazards fell in their way, one by one, to the peaks of the mountains. They would eventually get to the top.

Move forward to the early twentieth century and exploration by air. We're looking for the highest mountain peak in a range on which to site a radio beacon. So we drop a hundred paratroopers in a lot of places at random, each equipped with an accurate altimeter. We tell them to climb, and every week we check back to see how they are doing.

Some of our pathfinders have not got anywhere near a mountain and haven't climbed very much at all. We just give them a rendezvous point and tell them to go home. Some have made some progress, though. So now we drop in reinforcements. We don't put them in *exactly* the same places, but put in some little variations, just to try to speed up the search process. Another week goes by and another. Soon we've got nobody left on the plains, everybody is on a mountain. Some might have reached the top of minor peaks and can't go any higher. So they're being overtaken in terms of the altitude reached by others on the higher peaks. So we tell them to go home and we concentrate on the others.

With luck, after a few weeks, we'll have all of them on the higher peaks and soon some will reach the top. With luck, if we started with enough people, we'll have some on the highest mountain and they will eventually outstrip everybody else and find the highest point in the range. In fact there may be several peaks of very similar height. We might not get the absolute best. But we're still likely to find one peak good enough to site our radio beacon on.

This is the principle underlying evolutionary computation. It's a search process; we reward those that do well in the search and eliminate those that do not. The technical details in how we reward and eliminate are a major research frontier in computing at present. These *stochastic* algorithms—algorithms that contain a random component—seem to excel at finding very good solutions to messy problems, which like our mountain ranges have many irregular peaks and troughs.

We talk about a 'search space' for a problem, but what do we mean by it? Searching for a solution makes sense, but where do the peaks and troughs come in? We simply need a measure of how good any possible answer is. In our exploration example it was the height reached. In the Travelling Salesman Problem it was the distance travelled. In finding the optimal allocation of jobs to workers it might be the total idle time added up across all workers. In allocating rooms for lectures it might be how well the assignment of rooms fits into socially acceptable hours (no lectures before 10am, none after 4pm). These are just a few of the endless examples of complicated problems with lots of variables where evolutionary computation works well.

Most people will think of evolution as having something to do with genes or DNA and probably assume that it's linked to sexual reproduction. In evolutionary computation these factors all come in. They are the way we *encode* a problem solution. The genetic code gives us a particular type of eye—for cat, monkey, human—and one day we will understand exactly which genes correspond to what features of an eye. We already know, for example, what genes code for the coloured pigments in the eye's cone photoreceptors, and thus which defective genes lead to colour blindness.

Going back to our paratroopers, let's make the terrain a bit tougher: we'll add snakes, tigers, vicious bugs, steep cliff faces, deep rivers and foul weather. So to survive, everybody would carry guns, ropes, heavy boots, climbing gear, insect repellent—a pretty heavy mixture all up. It's expensive, too, so no single person has all the gear. Of course, the more you carry the slower you progress, but if you're missing a gun when a tiger pops out you might cease to progress at all. So now we add another strategy to our search. Every time we sort everybody into those who have ascended a reasonable distance, we also swap the equipment around. Eventually we end up with someone in the right place (the highest mountain) with the right gear to tackle the hazards *en route*.

In artificial evolution we invent a genetic code specifically for the problem we are trying to solve. The choice of this code is critical and finding good representations is a developing art. This art has been developing over some 30 years, but the theoretical paradigm has lagged somewhat. In fact we are in the rather unusual situation of having an intuitively appealing theorem crumbling under the weight of evidence, with no new theory to take its place.

Building blocks are vital to evolution. The flexibility for complete re-evolution of, say, an eye, is almost non-existent.[13] The vast complexity of living organisms comes from the continual reuse of components: proteins, cells, eyes, ears, limbs and so on. The *schema theorem* of John Holland, the founder of genetic algorithms research, encapsulated this idea of building blocks. Its intuitive appeal does not seem to match its applicability. Although it has not been disproved, the situations in which it works seem to be much more restricted than originally thought.

In the next chapter we shall think about the Internet and its nature as a vast testbed for artificial life. But before we leave evolutionary computation, we'll just briefly mention genetic programming, in which we evolve computer programs. Each gene in our artificial genetic code represents a token in a programming language. The whole genetic code represents a program. Most of the time this will be junk and the program will hardly run at all. Most random programs will be ineffective, but the search process will eventually find some that are functional.

Genetic programming successfully solves quite simple problems and solutions have been reported for more challenging ones. But at the time of writing, scaling up seems to be very difficult. However, we can see the enormous potential for the evolution of programs in cyberspace.

As we have talked of searching, we have implicitly assumed that the landscape is fixed and unchanging. We are looking for an optimal solution to a problem where the value of each solution is known, or can be calculated, in advance. But real biological systems are not like that. Species co-evolve and find environmental niches. Natural disasters take out food, species, habitat in sudden unexpected events. A much richer set of computational properties lie underneath such ecologies. We hope to see them in cyberspace real soon.

The immune system

Nature has another evolutionary scheme besides the one with which we are familiar in the replication of cells. The immune system has to recognise alien invaders in the body, and destroy and remove them. The killer cells generate templates for recognising specific invaders by an evolutionary mechanism, but there is no recombination of genetic material from parents in the traditional fashion. Everything is done by a process of mutation. Another important respect in which immune cells differ from a natural evolution is in their fitness requirements: there must be a huge population, with each cell adapted to some different invader, rather than all cells competing for some common fitness goal. In effect, each cell has its own environmental niche.

Some viruses, such as HIV, influenza and the common cold, are effective because they can rapidly change their protein coat. The immune system is fast and flexible at developing a response to a new invader. But it seems to be limited in plasticity: it mounts a new defence against *old* quarries.

157

Stephanie Forrest and colleagues at the University of New Mexico have been using the immune system as a model for computer security. The regular plagues of viruses that pass through the PC community cost countless hours of office time, while Internet attacks on enterprise computer systems are a major headache for system administrators. With the rapid growth in the Internet, computer *immunity* is becoming an urgent requirement!

So we have come full circle. From abstract models of computation we build digital computers. But then for some applications we find that simulating biological processes—nerve cell networks, the immune system, evolution—is the best strategy.

CHAPTER 9

THE INTERNET: SUPERHIGHWAY OR GOAT TRACK?

In 1991 US senator Al Gore (later Vice President) made a speech to the US Congress proposing development of a national research and education network. In this speech he referred to the creation of an electronic 'superhighway'. The term caught on. The next few years saw increasing hype in the popular media about the 'information superhighway'. Most of the interest has centred on the potential for video on demand, home shopping and other commercial innovations.

In practice the term 'information superhighway' at present means the Internet. Sceptics have pointed out that the Internet at the moment is more like a goat track than a superhighway. Despite this limitation the Internet is already changing the ways in which we do things in many areas of work and play.

Telecommunications in commerce is not new. Every time you use an automatic bank teller or make an airline booking you are using a computer network. What has excited speculation is the new possibilities created by the development of general-purpose, broadband data networks.

It is worthwhile looking at this phenomenon more closely. The Internet is an excellent example of dealing with complexity in practice. It also provides many examples of emergent properties—one of the hallmarks of complex systems.

WHAT IS THE INTERNET?

So just what is the Internet? Why has it excited so much interest? Behind all the hype we are witnessing an information explosion of historic proportions. The Internet has grown into a vast global communications network. Its origins date back to the 1970s when the US military was looking for ways to improve communication with its contractors. Out of this need arose ARPANET, which soon evolved into a national, and finally international, research network—the Internet.

The Internet is really an emergent phenomenon. Strictly speaking, it is not a network at all but rather a *protocol* for linking together many separate networks. In all there are over 10 000 of these networks, gateways and backbones. Between them they link together over five million individual computers around the world.

With so many computers the number of possible links is enormous—nearly 10^{14} combinations. The system copes with this complexity by organising machines into a hierarchy of *domains*. These are all linked via the Internet Protocol (IP). Every machine on the Internet has an *IP address*. This is a set of numbers that uniquely identify the machine and the various gateways that provide access to it. However, users rarely refer to these numbers directly. Instead they refer to names, which are easier to remember. For example, in the address 'life.csu.edu.au' the domain 'au' refers to Australia, the domain 'edu' refers to educational organisations, the domain 'csu' refers to Charles Sturt University, and 'life' is the name of a particular computer or service. Each domain in the hierarchy has servers that route messages to the appropriate domain or machine.

For most of its early life the Internet existed as a sort of anarchic club. Even among researchers only a few computer-literate souls were able to penetrate its secrets. They banded together to form the Internet Society, which still controls the Internet today.

A transformation came at the end of the 1980s with the invention of several new protocols. Protocols are simply ways of defining how one machine will talk with another. The most widely used protocols reflect the sorts of activities of the users. For computer programmers the most important activities were to communicate with one another (electronic mail), to access other machines (Telnet), and to exchange files (File Transfer Protocol or FTP). One important activity is anonymous FTP: many organisations provide public repositories of free software, images and data. Any Internet user can access this material via FTP, by using the name 'anonymous'. Surprisingly, the US military have maintained some of the biggest public FTP archives.

The importance of anonymous FTP was that it demonstrated the enormous potential of resource sharing. If you make your software freely available, then not only does it save others the time and effort of 'reinventing the wheel' but also they might help by adding to it.

Some examples illustrate this principle in action. The first is FRACTINT, which is one of the world's best fractal programs. Not only is FRACTINT freely available over the Internet but the entire source code is too. As a result, hundreds of people have helped to build and improve the program over a period of years. Recently, Chris Langton and Stephanie Forrest at the Santa Fe Institute have initiated software development projects— SWARM and ECHO respectively—using a similar approach. The SWARM software provides a kit for simulatiing complex systems, using a multi-agent approach. The ECHO project is developing a suite of tools for applying and using genetic algorithms.

However, by far the best examples of collaborative software are the GNU project and the development of

161

Linux. By the late 1980s the Unix operating system had become the most popular platform for research computing. The GNU free software project started at MIT with the intention of developing quality software for Unix operating systems. One of the most impressive products was the compiler GCC, which turns source code written in the language C into working programs. This program outdid its commercial competitors in almost every respect.

How was this possible? How could a piece of free software be better than the best commercial product? The answer is simply weight of numbers. A large software house might be able to put (say) a hundred programmers to work on a major piece of software. However, this is nothing compared to the forces that the Internet can muster. Try to imagine literally tens of thousands of enthusiastic 'hackers' scattered over the Internet and all vying to outdo one another to improve a piece of code. By the early 1990s the work of this group reached its logical conclusion with the development of Linux. Named after Linus Torvalds, the Finnish scientist who pulled it together, Linux is a complete Unix operating system in the public domain.

As the volume of free information grew, the problem of finding desired items grew ever more difficult. One of the first solutions was Archie, a system of directories named after a comic strip character. It was quickly realised that the process of looking up an entry in the Archie directory and then retrieving the listed item from the appropriate archive could be automated. This inno-vation pointed the way ahead to developments that triggered the current revolution.

WHAT IS THE WEB?

The need for an efficient way to transfer documents over the Internet is obvious. Many organisations devote a lot of time and effort to delivering information to people. There are huge potential cost savings in having interested people retrieve copies of documents, reports, announcements and so on for themselves.

Several problems plagued early attempts to develop an effective system for transferring documents over the Internet. For a start there was no universal standard for document formatting. Then there was the problem of how to index documents. But mostly there was the problem of developing a simple, automatic way of accessing, retrieving and displaying documents.

The Gopher protocol, developed at the University of Minnesota, offered one solution. In Gopher, the user encountered a series of menus. Each menu item was a link, either to a file or to another menu. The user had simply to click on an item and it would be retrieved automatically.

Gopher demonstrated the power of on-line document retrieval. Also, for the first time, it made on-line information accessible to users who were not computing experts. Almost overnight, a system that was developed to serve as a campus information service became an international information network.

Unfortunately, Gopher suffered from three major limitations. Firstly, its menu system was rigid and at times frustrating. Secondly, it did not prescribe a formatting standard for documents. Thirdly, indexes and documents were separated.

Meanwhile, researchers at CERN in Switzerland developed a different approach, the HyperText Transfer Protocol (HTTP), which overcame the above difficulties. In hypertext, words or phrases in a document can be made into 'hot links' to other documents. This means that we can incorporate Gopher-style menu items directly into a document. Moreover, HTTP prescribed a clear formatting standard as well—the HyperText Markup

Language (HTML). In computing terms, it operates via a client–server model. That is, a user's program (the *client*) sends a request to an HTTP *server*. That server then sends back the requested document to the client program, which displays it for the user.

The crucial feature of HTTP is that it provides links between on-line documents. Sets of documents that are linked in this way can be thought of as 'clusters'. As we saw in earlier chapters, the formation of linkages and clusters is at the heart of self-organisation in complex systems. So in a very real sense HTTP provides a mechanism that promotes self-organisation of information on the Internet.

Having developed HTTP, the CERN group set up an experimental set of sites for their system, which they called the World Wide Web. For a time the Web seemed doomed to oblivion for want of an effective interface. Then, in 1992, the US National Centre for Supercomputer Applications (NCSA) developed a brilliant program called MOSAIC that made the Web come alive. The number of users and publishing sites began to grow exponentially, and the rate increased markedly when browsers for PCs and Macintosh were developed. At one stage the number of users was doubling every nine weeks. By the mid-1990s Web activity had become the dominant traffic on the Internet. The World Wide Web is now firmly established as the Internet's publishing medium.

Many hands make light work

The development of information services on the Internet exhibits both self-organisation and emergent properties. We can best appreciate this from some examples.

The Internet is a complex system that is just beginning to organise itself. One organising trend arises from the need to be able to draw together information from many different sources about a particular subject or for a particular purpose. Many people and organisations compile databases of references about their area of interest. In some instances this activity can create new resources where the whole is quite literally more than the sum of its parts.

A good example is the Virtual Tourist. This information system allows users to zoom in from a map of the world to any country and call up basic information. Very quickly the user can be checking pictures of (say) the Botanic Gardens in Edinburgh or the current weather in Toronto. The point is that the Virtual Tourist is not managed by a single agency. It arose by systematically linking together information services from literally thousands of different Web sites all over the world.

Another emergent service on the Web is the Virtual Library. This is really a huge table of contents for the Web. Each topic in the directory, such as 'complexity systems' or 'computing', sends the user off to a server that maintains a comprehensive index of links about the subject concerned. Like the Virtual Tourist, the service integrates information from tens of thousands of separate Web sites.

The Internet is also influencing people's activities in areas away from the computer. The achievements of researchers in molecular biology provide an excellent example. In the late 1980s biologists set up several major databases on the Internet as repositories for biological data, especially gene and protein sequences. Prominent examples include Genbank in the United States and EMBL in Europe. It is now compulsory for researchers to submit their raw data to public databases such as these if they want to publish their research in any leading journal.

The present situation has arisen for a good reason. By adding together everybody's data we get more than we started with. Not only can scientists use the databases to help them interpret their own results but they can carry out entirely new kinds of study. For instance, they can search for patterns in the vast pool of data about DNA and proteins now available within these databases.

The above examples suggest ways in which information services on the Internet are likely to develop in the future. In most fields of activity there are enormous potential benefits in developing joint resources of the kinds described above. As a result we are already seeing many attempts to create 'information networks', 'virtual communities' and other cooperative projects. It is easy to identify many issues, such as documenting the world's biodiversity and studying global climate change, that call for joint efforts of the kind we already see in biotechnology.

The knowledge web

The Internet is perhaps the ultimate complex computer. At present, the bulk of traffic across the information superhighway consists of documents, images and electronic mail messages. However, linking computers together makes it possible to spread the burden of calculations across several machines. The Internet is ideally suited to distributed computing, where different machines specialise in solving different aspects of a problem. We are already seeing this phenomenon with the World Wide Web. What happens is that organisations publish information about their specialised area of knowledge. For example, the Weather Bureau publishes daily temperatures, the stockmarket publishes share prices, and so on. This means that we can draw together reliable, up-to-date information about a subject by consulting services provided by organisations that deal with each aspect of the problem. Suppose, for instance, that you want to compile a report about an endangered plant. You can obtain basic taxonomic information from the botanic gardens. National environmental agencies provide data about its distribution and about conservation measures. Universities and government research organisations provide details of current research about the plant.

The above sort of specialisation is now spreading to include many kinds of computation. For instance, in the course of their work scientists, engineers and other professionals often put together computer programs to carry out sophisticated and specialised tasks. In many cases the software would be of use to many other people, but the development of such programs is extremely costly and time-consuming. However, non-specialists may only need to use the particular calculation involved once. Therefore it is not worth their while buying an expensive piece of software for the purpose. As an alternative, organisations can now provide a user interface to their specialised software and services on the Internet.

In the above scenario, Internet computing is no different from what already happens in many industries where a project manager calls in a host of specialist contractors. However, by automating and standardising the process we are likely to see Internet computing go much further. Suppose that a query system does not need a user to sit down and fill in details of the processing required. Suppose instead that it accepts the job as a bundled query 'object' from another computer, and that it sends back a result 'object' in the same way. Then we could write programs that include automatic queries to other machines.

All of the above possibilities are already being explored. However, if the theory is straightforward, the practice is not. At present, the Internet is a disorganised mess. While it is true that we can now combine information from many different sources, locating that information is a huge problem. At the moment, the most common way of locating information is via one of the main Internet search engines. These are huge databases that list the contents of thousands of sites. Sometimes this can be a useful way of discovering previously unknown resources. However, it can also be extremely frustrating. For example, there are tens of thousands of references to Shakespeare on the Web; and if you want to look up information about flu and other bodily viruses then you are likely to be swamped with references to computer viruses.

We have alluded to two processes that are promoting organisation. These act as filters, in which similar information resources are linked together. Likewise, there is the need to be able to draw together information from many different sources about a particular topic or for a particular purpose. So people and organisations are compiling databases of references about their field of interest.

One of the driving forces behind the present anarchy on the World Wide Web is the increasing commercial activity in which companies see the Web as a means, not for sharing information, but for promotion and advertising. So besides promoting an intense aversion to coordinating and sharing on the part of information providers, it has led to an incredible duplication of effort. The reason is that the quickest way to promote your Web site is to provide material that will attract users, even if that material is already provided elsewhere.

For instance, one of the services that the authors provide on the Internet is a 'Guide to Australia'. This is a distilled index to essential information resources about Australia. However, over the past two years almost every service provider that has set up on the Internet has duplicated this service in one form or another. Despite the current emphasis on sites, rather than services, there is great pressure from many directions to organise information on the Web. One of these pressures comes from the need for organisations to coordinate their activities. Libraries, for instance, need to share information about their holdings—for handling interlibrary loans and so forth. Similarly, many government and scientific organisations need to share essential data and resources. This has led to the spontaneous formation of many professional information networks, virtual communities and other similar mechanisms for coordinating the provision of information. In time, hopefully, the present preoccupation with individual sites and 'home pages' will decline and topics and activities will become the focus of organisation. In the coming century the above processes should see the Internet grow into a comprehensive, worldwide information system. In the scientific arena we can see a clear historical progression in the way discoveries are communicated. In the time of Newton and Galileo letters were enough to keep colleagues informed of new results. By the nineteenth century we had the age of the great scientific societies, with discoveries announced in formal lectures to the Royal Society. The twentieth century saw the heyday of the scientific journal. The twenty-first century will surely be the age of what we might describe as the 'knowledge web'.

All of the above sounds absolutely thrilling and there is reason to expect that the Internet will improve our lives and make many new activities possible. However, there is also a darker side to the current information explosion. On the Internet we are already beginning to see the emergence of what can only be described as 'information imperialism'. First of all, the Internet uses English as its standard language. This is reinforcing the position of English as the prime language for international communication. It is a harsh reality that anyone whose Internet site provides a service in any language other than English will simply not attract a large international audience. Secondly, the Internet is serving to promote American ideas and culture. This is a function of the vastly greater number of sites and services emanating from the United States than anywhere else.

Finally, and perhaps of greatest concern, the Internet is tending to deny people a voice. This may sound a bit surprising at first. After all, on the Internet everyone is connected to everyone else. It is true that the Internet can level the playing field in some respects. Even a 'backyard' server has as loud a voice as the largest corporation. However, the problem is really whether people will look at what you have to offer.

Here we have the problem that network speeds are generally much faster in the United States than in many other parts of the world. Hence users expect very fast response times. They become impatient when services are slow to respond. This raises a problem for sites outside the US because international links are generally heavily used and generally slower than local links. So even though many countries have very comprehensive information and up-to-date and easy-to-use services about themselves, American users prefer to get the story from an American source. This situation is compounded by indexing sites pointing to those US sites, rather than the more authoritative and up-to-date sites overseas.

This aspect of the Internet reflects a similar process that is also under way in many other aspects of modern life, wherein the diversity of business and cultural resources is being overtaken by monolithic international corporations.

CYBERSPACE

You do not have to be a good scientist to write good science fiction. In a fast-changing field—and computers are probably the *fastest*-changing field at the moment—

knowing too many of the details can cripple the imagination. William Gibson, inventor of the term *cyberspace*, was not short on creativity. His characters inhabit twilight worlds of physical reality and an information space, in fact something like the Internet. But whereas Web servers are static and unintelligent, in cyberspace there is not only information but also intelligent beings.

Gibson's cyberpunks connect to the 'Internet' through direct brain implants. Farfetched? At present, maybe, but not outside the realm of imagination. We know, for example, that the cortex of the brain is at birth just a general purpose neural computer, which can develop in lots of ways. As noted earlier, one of the more intriguing (and maybe rather gruesome) experiments of recent years has been the demonstration that one can mix up cortical development.

Suppose that you surgically remove the optic nerve and plant it somewhere else in the cortex, say in the area destined to become the hearing centre. The amazing thing is that that part of the cortex now develops as a perfectly functioning *visual* area. In other words, what the particular piece of cortex does depends upon the input it receives. Now, the cortex has evolved expecting a visual input *somewhere*. One of the big questions in neuroscience at the moment is in what way evolution has prepared the brain to expect particular sorts of input. One definite possibility is that the main predisposition is towards recognising *structured* information—that it doesn't really matter what the structure is (whether it's visual, auditory, tactile or any other sort of information) as long as it contains patterns.

In this eventuality, the way is open for *cyberdecks*, where the brain plugs directly into cyberspace. It seems unlikely, though, that this would ever work for an adult brain. Thus the ethical conflict for experiments of this kind would never be resolved. The operations would

171

have to be carried out at a very early age, before the child had any vote as to whether he or she wished to be a cyberpunk!

Intelligent agents

So how many years are we away from the cyberpunk era? Perhaps not as many as we might think. Let's track the initial stages of the evolution of Wintermute,[14] the giant Artificial Intelligence in *Neuromancer*, one of the first Gibson cyberspace novels. The evolutionary force to begin with is the enormous but chaotic growth of the web. Lots of people have trouble reading all the email they are sent. Push technologies, which dump more and more information on the desktop of your computer, are already appearing. So we need to build filters, clever filters which know what we like and don't like.

Companies such as Firefly in the United States already build profiling agents. We can do a lot with simple statistical matching. For example, word frequencies are amazingly discriminating in characterising author style, so much so that experts are willing to stake money on whether Shakespeare did or did not write particular plays and so on. It just turns out that we tend to use some words more often than others (I'm particularly bad with 'particular'). So we could make a list of keywords that describe our interests and then look for frequencies of these words in email messages or web pages. This is what we might call a strong AI approach. We're defining an algorithm (a statistical formula and a data set) to do the job.

The above agents can be quite successful. But we can expect different things down the track. One alternative way to build an agent is to make its parameters *evolve* : so we might have a formula which determines how we do a match between the things we write on our own Web site and the things on other sites. Our agents could evolve this formula to make it more and more useful.

It would work something like this. When we do an Internet search, for example, we would be given an agent (that's one example of the formula) which would go away and do the search. Then when it returns we would give it a ranking, say by clicking on a score. A daemon sitting in the background would continually be mating agents against one another according to their scores, in a 'survival of the fittest' setting.

But we saw earlier that genetic programming is a reality: we can actually evolve computer programs out of random strings! So why not go one step further? Instead of specifying a formula at all, simply evolve the method the agent uses from scratch. We might need a lot of patience at the beginning, but it could pay off in that we might find solutions we never dreamt of. Although this seems to be suggesting that evolution can find better algorithms than we can create ourselves, this does in fact happen. Evolutionary researchers frequently see their programs finding unexpected solutions, often because they overlooked something in their experimental design—just like the undoing of Clever Hans.[15]

So far we are still in the present. All these sorts of ideas can and are being implemented here and now. But how do our agents get real autonomy? The secret has to lie in making the criteria for success, the fitness criteria, more and more general. It's like the difference between a secretary from the temp. staff agency and a personal assistant who has been with you for years. The instructions are fewer, the feedback is on a much longer time scale and there's a vast difference in the usefulness of the two. But in allowing the fitness criteria to become very general we are allowing our agents to evolve in perhaps unexpected ways.

173

The final stage is defining protocols on the Internet which enable agents to query one another. These automatic agents extend the idea of query objects, which we discussed earlier. We already have so-called 'common object busses' and object request brokers which enable computing objects to talk to one another, even though they may be on different machines of different architectures and be written in completely different programming languages.

Remember that the idea of an object is a chunk of data which is protected from the outside world and accessible only through special procedures, or methods, that belong to the object. In response to a message from outside, a method will return a piece of data, perhaps customised in some way.

The stage is set for the evolution of communication and independent action of agents and objects on the Internet. They can become a self-perpetuating computational ecosystem. From that point on, we lose control. Our only option would be destruction, but that is not likely to happen. The growth of intelligent agents will be fuelled simply by their usefulness. We will be reluctant to let them go.

CHAPTER 10
COMPLEXITY AND LIFE

What will industry be like a hundred years from now? It is sobering to try to guess how science and technology will develop in the future. Almost every area of technology has advanced dramatically during the twentieth century. It is reasonable to assume that progress will continue. However, if the past is any guide, much of that progress will be the replacement of one technology by another.

We have also seen the rise of entirely new technologies. At the turn of the nineteenth to the twentieth century, motorised transport gradually replaced horse-drawn vehicles. The twentieth century has seen the rise of electronics, computing and aeronautics, to name but a few new industries. At the century's close many more technologies lie on the horizon. For example, although it is already very active, biotechnology is poised to spawn major industries during the twenty-first century. Likewise, space travel, and a host of possible technologies that exploit micro-gravity, are still in infancy. In the context of this book, the question of greatest interest is what new discoveries and what new technologies might stem from the development of complexity theory. As we have seen, discoveries about the nature of complexity in general have already led to new insights about specific systems and processes. Studies of chaos are helping

175

scientists to understand the nature of turbulence. No doubt there are many more discoveries to come.

When we consider technology, complexity has a lot to offer. Perhaps the greatest prospect is the possibility of learning how to control processes that involve really complex interactions. Take biotechnology, for instance. The essence of biotechnology is to duplicate organic processes in an artificial environment. The great breakthrough offered by genetic recombination is that it is now possible to manipulate and adapt natural processes to our own ends. Impressive as they are, the current generation of applications are really fairly simple. A good example is the production of interferon, a substance that is produced naturally by the body in small quantities. It plays an important role in helping the body to ward off disease. Almost as soon as it was discovered, the potential value of interferon in medicine was realised. The problem was the extreme difficulty of extracting enough for medical use. Genetic recombination solved the problem. By splicing the fragment of human DNA that coded for interferon into bacteria, the bacteria were persuaded to manufacture interferon as a by-product of their metabolism.

The above one-to-one correspondence of spliced genes to single products can only take biotechnology so far. A much greater challenge is to learn to emulate genetic control over development. Most stages of development and growth require ordered interactions between many genes. The work of Kauffman and others is beginning to reveal how growth processes are organised at the genetic level. The ultimate possibilities that a full understanding of growth would lead to are simply breathtaking. Many machines will contain organic components. Many materials, and even complete products, could be grown instead of manufactured.

176

UTOPIA OR BRAVE NEW WORLD?

In the above discussion we have looked at just some of the possibilities opened up by new technology in the coming century. However, we should stress that our discussion is definitely not meant to be an argument for blindly embracing new technology. Recent developments in medical technology—cloning, embryo transplants and the like—have brought into sharp focus the question of whether or not we should adopt new technologies merely because they are possible. Any new technology has two sides. On the one hand it solves problems or creates new opportunities. On the other hand it also has an impact on the *status quo*. Its effects can be economic and social, as well as technical. Moreover, these effects may be negative. So any benefits that a new technology conveys should be weighed up against the costs. New technologies have often had undesirable side effects. Let's look briefly at some examples.

The invention of television was one of the great achievements in communications during the twentieth century. Although TV has provided great benefits, it has also had a huge impact on society. In the Western world it is arguably a major factor in alienation and social breakdown. In Third World countries it has promoted a widespread partiality to Western culture, disrupting local customs and tradition. Improvements in transport have a similarly chequered history. In Western cities fast, cheap public transport has led to the development of vast suburban dormitories. The average standard of living has increased. However, like TV, transport has also contributed to social breakdown. People move further and more frequently. Friends and family often live far away. People have less attachment to their local community and fewer social contacts.

177

In industry, automation, such as the introduction of robots, has increased production and helped to greatly reduce the costs of products to consumers. But these advantages have also contributed to enormous cuts in employment in the industries concerned. Perhaps the greatest advance of all has been the enormous improvement in public health. In most parts of the world, life expectancy has increased and the traumas of disease and death in childhood have been greatly reduced. Great though these benefits undoubtedly are, they also have a negative effect: overpopulation. For almost a century now the world has been suffering from runaway population growth. The results are widespread poverty and suffering, plus periodic incidents of famine and starvation.

Some of the worst stories arise from mismanagement of the environment. We simply do not know enough to get it right. Earlier we saw the example of the cane toad in North Queensland. Environmental management is full of stories of misguided schemes that went wrong. Genetic engineering raises the spectre of even greater environmental disasters. The fear is that genetically altering (say) a virus to control some pest might produce a supervirus that attacks other organisms, and even humans.

The recent story of the calicivirus illustrates just how potentially disastrous genetic manipulation could be. Rabbits have been a major environmental problem in Australia for over a century. For many years the disease myxomatosis kept numbers in check. But by the 1990s the rabbits had become resistant and their numbers began to increase again. To replace it the CSIRO[16] genetically altered a calicivirus. Aware of the dangers, they carried out an experimental release into a rabbit population on an island off the southern coast. Almost at once the virus mysteriously started turning up in rabbits on the mainland. No one knows how it got there, though desperate farmers were suspected. Within weeks the virus had

178

spread over a vast area of the mainland rabbit population. Fortunately, there seem to be no obvious side effects. However, the risk is that one day such a virus might cross over into other species and wipe them out too.

THE ANATOMY OF AN ACCIDENT

It is a dark and stormy night. Rain is teeming down. At a busy highway intersection a driver who is hurrying home starts to make a turn. Out of nowhere another car suddenly appears. It is coming straight for him. Both drivers slam on their brakes, but cannot stop in the wet conditions. There is a sickening crunch and both drivers end up in hospital with serious injuries.

Why did the accident occur? Whose fault was it? Such questions are often hotly disputed. They can be the subject of much passion, anguish and litigation.

The truth is that accidents are complex events. They often result from a deadly interaction of many factors. One popular saying is that accidents happen when things are different for one reason or another. In the above example, the storm is a major difference from normal driving conditions. So is the pressure on at least one driver to get home quickly.

In general, accidents occur through the interaction of different factors. For example, everyone has a *reaction time* —the time it takes to react to a new situation. Potentially dangerous situations, such as cars pulling out into our path, occur all the time. Normally our reaction time is perfectly adequate to cope—otherwise driving would be impossible. However, many factors can affect our reaction time. For instance, tailgating, where one car is moving along very close to the one in front, can reduce the available time to react to virtually zero. So can poor light and wet roads. Also other factors, such as distraction or fatigue, can slow down the driver's reactions.

179

Any combination of factors makes for a very dangerous situation indeed. We may be able to cope with (say) wet conditions most of the time. But in extreme situations, where we need to perform best, we are most likely to fail. And these extreme conditions are more likely to occur when other factors, such as bad roads or driver fatigue, enter the equation as well.

Deadly combinations of factors are well known to air crash investigators. Sometimes a chain of events builds up, with fatal results. For example, the crash of an aircraft into the Potomac River, Washington DC, in the mid-1990s involved delays in takeoff, due to snow and ice. This put pressure on the flight crew. After first forgetting to turn on the de-icing equipment, they rushed through their checklist and failed to act on the error. Instead, they tried to take off anyway.

Accidents in the home often build up in a similarly tragic way. In a mad rush to get off to a party, Mum and Dad leave electric hairdryers and razors connected and lying on the bathroom bench. The babysitter, being a stranger, assumes that all is safe when she sends junior off to the bath. But the toddler, finding a new toy, plunges the hairdryer into the bath and is promptly electrocuted.

The events leading up to an accident can be long and complex. They can involve the interaction of many small events and circumstances, which by themselves are not significant. People like things to be simple. We always try to identify simple patterns of cause and effect. So the complex nature of accidents can be both confusing and frustrating.

In trying to understand the nature of cause and effect, philosophers used to talk about 'immediate' and 'final' causes. Thus, in the above example, plunging the hairdryer into the bath is an immediate cause of the accident and leaving the hairdryer turned on is a final cause. However, as we have already seen, there can be many contributing factors, so even this categorisation is too limiting.

None of this is helpful to the relatives of victims, as they try to come to terms with their loss. There is a tendency to pin blame, to seize on one of the many factors that may have been involved and present it as 'the cause'. It is all too easy to put the blame on a dead pilot, for instance.

Lawyers, with an eye on larger fees, are all too easily tempted to place the blame for an accident wherever there is a potential for a payout to their client. In the United States, this habit has led to some unfortunate and bizarre results.

One result is a tendency to sue doctors the moment anything goes wrong. The factors contributing to illness and death can be just as convoluted as accidents. The idea of malpractice is to guard against gross negligence, such as sawing off the wrong leg. But where does negligence end? In the extreme, one can argue that there is always something more that can be done to save a dying patient. So it can be argued that any time a patient dies the doctors are *ipso facto* negligent. Huge payouts in malpractice suits seem to be commonplace, especially in the US. The fear of litigation affects the way in which doctors work. For instance, obstetricians who are anxious to avoid complications have become increasingly inter-ventionist. One sign of this is an increase in the frequency of Caesarean sections.

The risk of litigation also leads to a feedback effect in which higher payouts force higher insurance fees for doctors. Doctors have to pass on those costs in the form of higher fees for service and their patients therefore have to pay much more for medical insurance. Poorer families cannot afford the rates, and avoid visits to the

181

doctor. Their standard of health goes down and death rates go up.

Litigation related to accidents has reached bizarre levels. If a burglar trips over a rake in your backyard, he can sue for damages. But more worrying still is a disturbing trend that emerges from accident payouts. Just as with medical malpractice, individuals and companies are liable if gross negligence leads to injury or death. But again, what is negligence? Is it negligence if a customer rushing out of a store slips on paving wet from the rain and injures her back? Should the company have anticipated that she would have been rushing and put up signs urging caution?

How far should this sort of thing go? What sorts of dangers can an organisation reasonably be expected to anticipate? If a car crashes through the front of a shop, is the shopkeeper liable for damages to customers because he or she failed to install safety glass in the shop window? If a diner at a restaurant gets sick from overeating, is the restaurant at fault because it provided the food? If a man injures his back in a novelty contest where entrants have to run a race carrying a refrigerator on their backs, is the manufacturer at fault because it failed to warn the man that such activity could be dangerous?

What we are seeing here is a gradual process in which doctors, airlines, manufacturers—in fact everyone providing any sort of service at all—are expected to assume more and more responsibility for the health and safety of the population. Conversely, the sorts of cases we have seen above contribute to a culture in which individuals are encouraged to assume no responsibility whatever for their actions.

All of this contributes to an increasingly consumerist culture in which individuals expect everything, including care, to be provided for them. Consumers assume no responsibility at all. One could reasonably argue that it is a civic duty to alert fellow citizens to dangers. If a shopper falls over on wet pavement, why not sue the passers-by who failed to alert her about the slippery pavement?

In hindsight, almost any accident is avoidable. The underlying problem is one of complexity. There are always so many combinations of factors that we can never anticipate them all. Also, some dangers can be anticipated but are so rare that to try to counteract them all would be prohibitive in time and effort. For instance, on any suburban street there is always a chance that somewhere, sometime, a child will dash out from behind a parked car and be hit by a passing motorist. This does happen. Should every street carry large warnings? Should we restrict speed limits to 5 km/h to ensure that drivers can react?

This brings us to the general issue of prevention and precaution. Some dangers are obvious and real. Building codes include measures to reduce the risk of fire. Likewise there are codes for occupational health and safety, for food preparation, and for a hundred and one other things. But how many precautions can we take? To take an extreme case, what should we do about the danger of meteorites? People have been killed by falling meteorites. Anytime, anywhere, there is a finite danger that a meteorite will strike. It may be small, but it is real.

One issue that modern culture has not dealt with very well is balancing risk with reaction. We've all heard stories about officials failing to install crossings or take some other action on safety until someone has been killed. Conversely, there are many examples of over-reaction. Air hijacks are extremely rare events, compared to the total amount of air travel. Nevertheless, they attract such vast media attention that airports everywhere now spend millions of dollars every year on security.

Dealing with accidents has played a surprisingly large role in shaping society. Modern society has many institutions that deal with accidents in one way or another. Examples of mechanisms to prevent accidents include traffic laws, building codes, regulations for food preparation and for occupational health and safety, and drug testing. And there are the institutions that respond to accidents, including ambulance, fire brigade, search and rescue units, and police.

Just as the increase in complexity of our society brings with it increased risk of accidents, so the change in managerialist culture brings with it hidden dangers. Biological systems are highly adaptive, but they also possess considerable *redundancy*. Engineers have understood this for a long time. You can't expect a message to be transmitted across a communication channel perfectly all of the time. So what you do is to build in extra information contrived in such a way that if any particular bits are lost, it is still possible to work out what they were. In the times of massive over-employment we had considerable safety checks built in to organisations simply because there were enough people around with enough shared knowledge to catch other people's mistakes before they became life-threatening. The move to ever greater efficiency has cut out much of this safety net. A tragic example was the *Herald of Enterprise* disaster, in which a sea-going ferry sank, drowning dozens of passengers. This occurred because one person had ultimate control of the ferry doors and, one night, he accidentally left them open. If we have to make organisations more efficient, we also have to redesign the safety nets. *Until we fully understand the dynamics of complex systems we may not be able to do this properly.*

In the above stories we have tried to provide examples that show how *complex* phenomena such as accidents influence society. The immediate effects can be subtle, and perhaps minor, and yet the end results can be profound changes in our social fabric.

SYNCHRONISING SOCIETY

How many people do you know? For most people the answer would run to scores of friends, family, colleagues and acquaintances. If you count everyone you ever met, then the answer could easily run to hundreds of people. However, if we restrict the count to individuals in regular (if infrequent) contact, then the number rarely goes beyond 200.

In Chapter 6 we saw how interactions between people define connections and networks within society. Just how influential are networks in shaping human culture and society? To look at just one type of network, we can ask what role gossip plays in human society. The instant response is almost invariably to assume that gossip is a trivial activity. However, Robin Dunbar, Professor of Psychology at Liverpool, suggests that gossip has played a crucial role, not just in the history of human society but also in the evolution of language and of humanity itself.

Gossip is the everyday exchange of ideas and pleasantries. We associate it especially with the spread of stories about social matters. Dunbar suggests that gossip evolved in early humans as a kind of verbal grooming. Now, it is well known that grooming in ape societies is an important social activity. It helps to maintain social cohesion. Dunbar's theory is that, at some point in human evolution, speech began to take over the role that grooming plays in troops of apes and monkeys. He goes on to suggest that this new role was an important factor in the development of speech and hence in the evolution of *Homo sapiens* as a species.

As a form of grooming, speech has the advantage that you can talk to several people at once. The one-on-one grooming seen in (say) baboons or chimps is slow. To interact with every member of your troop by grooming takes a long time because you have to groom them one at a time. However, if you can talk, then you can interact with up to three or four other individuals at one time.

Suppose that you're talking socially with a few friends at a party. Then if the group is small (less than five), everyone joins in the conversation. But if a couple more people join the group, it quickly breaks up into two separate groups, each discussing different topics. This sort of phenomenon suggests that there is a natural maximum group size (about four) for social discussion.

Now if we suppose that gossip is social grooming then we should find that the size of human clans is larger than ape and monkey troops. It is well documented that apes and monkeys form troops of about 30–50 animals. Dunbar has gathered a lot of evidence to suggest that there is a natural troop size for humans of about 100–150 people. He cites examples such as the Roman *century* and the fragmentation of religions into communities as cases in point.

This entertaining concept has a neat physiological twist to it. Why should grooming bind animals together? It transpires that grooming not only serves a useful function, in, say, removing twigs and burrs from the fur, but it also does something unexpected. It causes a release of the natural opiates in the brain, creating a pleasurable experience. It appears that laughter also releases natural opiates, thus allowing language to bind animals together in the same way as physical grooming, but on a larger scale.

Gossip does play an important role in shaping community attitudes, especially (say) in small towns or other groupings. Conversely, the same theory would argue that lack of communication is a major contributor to the problems of alienation and antisocial behaviour that we see in many Western cities. The erosion of traditional religions and the rise of TV culture have presumably contributed to this process.

As far back as we can trace, people seem to have had religious beliefs. These beliefs have exerted an incredibly powerful hold on individuals. Saints have suffered uncountable horrors, Abraham would have sacrificed his son, while hundreds engaged in mass suicide in the Jonestown religious cult. Do we learn anything from complexity about how religion can produce coherent behaviour among human beings? We might suggest, tongue in cheek, that Prophets embody the slaving principle (see Chapter 3). They convey a message so powerful that others follow regardless of their own desires. Other religions have rituals, varying in frequency and depth, which provide the synchronisation keeping groups together.

But it is internal codes of action, be they religious or otherwise, that can produce distinctive societies. Changing the rules and extent of interaction will produce just as strong a social pattern as rigorous central control. But as with the other complex systems we have seen throughout the book, working back from desired pattern to rules is far from trivial, if not impossible.

LIFE THROUGHOUT THE UNIVERSE

The year was 1976. After a journey of over 150 million miles the first Viking probe was about to land on Mars. Interest was intense. For the first time ever a probe had been sent to another world with the specific aim of searching for signs of life. Both scientists and an interested public eagerly awaited the first photos. What would they show? Would we see actual living creatures from outer space? Would science fiction at long last become science fact?

The history of interest in life in outer space goes back a long way. Ever since astronomers first became aware that the planets were really other worlds like the Earth, people have speculated about whether or not life (and particularly intelligent life) exists elsewhere in the universe. The expectations of finding life on Mars took a severe battering in 1965 when the Mariner 4 space probe swung past Mars. The first close-up photos of the planet showed a barren surface pockmarked with craters. Mars suddenly seemed more like the Moon than the Earth. Instead of little green men, microbes seemed to offer the best hope of finding life on the red planet.

This, then, was the background to the Viking landings in 1976. When the first of two landers touched down on the surface, no one really expected Martians to come marching up to the machine. There was still a real hope of seeing lichens and other primitive plants. But there was little surprise when the first photos showed a barren, rock-strewn landscape. At least the sky was pink.

Much more faith rested with the lander's shovel, which scooped up a sample of Martian soil for analysis in a robot laboratory. The scientists planning the Viking mission had devised what they thought would be a series of definitive experiments to test conclusively whether organisms lurked in the Martian soil. These experiments included tests for three of the main characteristics of biological activity (on Earth)—respiration, growth and metabolism. In what is perhaps the best tradition of science the results confounded everyone. Two of the experiments gave negative results, but one was positive. No one had anticipated anything like this. Something very strange was happening in the Martian soil. The

Viking experiments prompted a lot of fresh thinking about what was going on. One theory was that perhaps the Martian soil was like soil on the early Earth. On Earth today, air, water and soil are almost everywhere affected by the action of living things. Perhaps on Mars there were certain chemical reactions that on Earth were once quite common but have been obliterated by the activity of living organisms. All the same, the essentially negative findings of the Viking mission all but ruled out Mars as a serious candidate for hosting life. Attention therefore shifted to other possibilities, both within the solar system and beyond. Associated with this shift was an increase in interest in the origins of life.

For a long time the abundance of carbon dioxide and nitrogen in the atmospheres of Mars and Venus was taken as evidence of their similarity to the Earth. The assumption was that in the early solar system all three planets had atmospheres that resembled those of the outer planets. The outer planets have an abundance of methane and ammonia in their atmospheres. The theory was that the activity of living things had transformed the atmospheres of the inner planets to give them their current composition.

In 1953 Harold Urey at Chicago University performed an elegant experiment to test this theory. Into a set of connected flasks he placed water in an atmosphere of ammonia and methane. He then passed electric sparks through the gases of the flask to simulate the effect that lightning might have had in the early atmosphere. After a week the water had turned into a pink soup laced with a wide variety of organic compounds. These compounds included amino acids, the building blocks of life.

These results provided a perfectly consistent scenario for the origin of life on Earth. Over millions of years lightning storms in the atmosphere would have produced vast reservoirs of organic matter, creating an environment apparently ripe for the appearance of life. Now that probes have carried out close encounters with most of the planets, it is apparent that differences in their composition and atmospheres are linked to the origins of the solar system. In the condensing clouds of gas and dust that formed the early solar system the inner planets accumulated a very high proportion of heavier elements, whereas the outer planets acquired high proportions of gases. As a result the original atmospheres of Venus, Earth and Mars were quite different from those of Jupiter, Saturn, Uranus and Neptune. Current thinking is that the atmospheres of Venus and Mars today retain essentially the same composition as they had when they formed. The Earth, however, started with an atmosphere very similar to those of the other two, but living processes have transformed it to give the present composition. It is now clear that the clouds of Venus are the end result of a runaway greenhouse effect—and a dramatic example of positive feedback in action. Venus receives nearly twice as much radiation from the sun as the Earth does, but the dense clouds reflect much of it back into space. That is why Venus is the brightest 'star' in the night sky. However, the telling factor is that the dense atmosphere and heavy cloud cover retain heat, allowing the surface temperature to climb to a hellish 800 degrees Celsius.

There is an important lesson in this. Here is a planet that is virtually the twin of the Earth and there is every reason to suspect that Venus might originally have had all the ingredients that led to the appearance of life on Earth. And yet the Earth is ideal for life, whereas Venus would fry any living thing in seconds. Perhaps this is a cosmic lesson that we should heed. It teaches us that the greenhouse effect is not a figment of some alarmist imagination. Unless we act there is a very real risk that global warming arising from the build-up of greenhouse gases in the Earth's atmosphere will become irreversible. In the worst case scenario the Earth could literally end up like Venus and life, all life, on this planet would end!

LIFE AND ITS ORIGINS

To understand the origin of life we first have to consider the underlying chemistry. It begins with mechanisms that *enhance*, not suppress, order and pattern. Ilya Prigogine was concerned with the way in which self-organisation occurs in physical systems. He focused particularly on *dissipative systems*. These are systems that produce energy internally and exchange it with an external environment. They include living organisms and many chemical reactions.

Prigogine pointed out that, in dissipative systems, minor irregularities can grow into large-scale patterns. That is, such systems are highly sensitive to random variations. This means that in a prebiotic world dissipative systems would have provided ideal conditions in which novel chemical structures could appear, spread and diversify. An environment that suppressed chemical variation would have prevented complex organic compounds from appearing at all.

Organisms must have a source of energy. On Earth the two main processes are photosynthesis, in which plants trap light energy from the sun, and respiration, in which cells 'burn' carbohydrates and fats. The assumption has generally been that life elsewhere in the universe must use the same processes as we see here. However, the discovery of organisms that metabolise sulphur and rely on the heat from volcanic vents shows that living things can adapt to any available source of energy. Also, it is now understood that today's oxygen-rich atmosphere on Earth is a relatively recent phenomenon. The first living things on Earth survived in the absence of free oxygen. The appearance of the first plants, which release oxygen as a by-product of photosynthesis, probably created an environmental crisis for other organisms.

The large gas giants of the solar system—Jupiter, Saturn, Uranus and Neptune—have usually been ruled out of consideration as candidate worlds for life elsewhere in the solar system. One reason for this is that they have reducing atmospheres, instead of oxidising. Another is that there are relatively small amounts of water. However, there may be chemical alternatives. Ammonia, for instance, has chemical similarities to water.

What are the fundamental requirements for life to thrive? We need to look for several criteria. Firstly, there has to be an adequate energy source. For the Earth, this source is solar radiation. Secondly, there has to be an environment that lies within an acceptable temperature range. Although the Earth has extremes, most of the world most of the time lies within the range of 0–30 degrees Celsius. At very low temperatures, chemical reactions slow down and even stop. At high temperatures, proteins become denatured; that is, their delicate three-dimensional structure is distorted as chemical changes occur along the protein molecule. This is a problem on Mars, where even during the day temperatures can remain far below zero. Many worlds have zones that lie within a suitable range, at least for part of the time.

Thirdly, there needs to be an abundance of essential chemical elements from which to build bodies. It is almost certainly no accident that life on Earth is based on carbon. The chemical properties of carbon make it ideal for building complex molecules of the kind needed to evolve living things.

Finally, and most problematic, there needs to be a medium rich enough in interactions and variety to allow chemical and biological evolution to proceed. One problem with the idea of organisms in the Martian soil is that the medium provides very low mobility. In contrast, the Earth has an abundance of mechanisms. There is vigorous debate over the medium in which life may first have appeared on Earth. One view is that the oceans were originally a soup of organic compounds within which chemical interactions eventually gave rise to self-replicating forms. Other scientists claim that the oceans could not have achieved a sufficient concentration of the necessary chemicals for chemical evolution to occur. Shallow water pools in which organic compounds had become concentrated would have been far richer in chemical interactions.

Perhaps the best candidate for a home for life in the solar system is the planet Jupiter. It satisfies all the criteria listed above. Solar radiation is much weaker on Jupiter—even than Mars. However, Jupiter is so massive that it is really a failed star. Its internal heat is so great that it actually radiates about four times as much energy as it receives. Temperatures at the top of the atmosphere are well below freezing, but temperatures at the centre of the planet are very high. Somewhere between these two extremes is a zone where the temperature is identical to that of the surface of the Earth. However, there is a complication. The composition of Jupiter is so different from the Earth's that it's not clear whether Jupiter has a surface at all. If it does, are there oceans of any kind? And do temperatures in the right range occur where organic chemicals can concentrate and interact?

On the other hand, evidence suggests that Jupiter has the right 'laboratory conditions' for creating the ingredients of life. Lightning has been observed in the Jovian atmosphere. Remember that the experiments by Harold Urey and others were based essentially on conditions found on Jupiter. So we can be confident that on Jupiter, where the experiment has been in progress for billions of years, there must be considerable quantities of amino acids and other building blocks of life.

To understand the origins of life we have to go back to the chemical processes that preceded the appearance of living things. How did life-like properties emerge during this period of prebiotic evolution? Perhaps the most basic characteristic is self-replication. It is generally assumed that the precursors to living things were molecules that were capable of duplicating themselves. Another relevant property is the ability to accumulate information by growing longer and longer.

Given the universal nature of genetic information in modern organisms, it is generally accepted that the chemistry of the genetic code must have been an important development during prebiotic times. The obvious candidate is DNA. But the mechanics of DNA replication are relatively complex. It looks increasingly as though DNA emerged later as an intermediate step in the replication of various forms of RNA. That is, it looks as though living organisms emerged in an RNA world as a device to assist in the chemical process of replicating RNA.

We are still a long way from understanding in detail the progression of prebiotic evolution. However, along with self-replication was the need for a mechanism that could accumulate information by accommodating longer and longer molecules. Here there is a fundamental difficulty. The problem is one of complexity. No process of information transfer is ever perfect. There is always a certain error rate. Now the problem, as Eigen and Schuster pointed out, is that self-replication is limited by an 'error threshold'. That is, for any given error rate there is an upper limit on the size of molecule. If the molecule grows beyond the critical size, then there is a cascade of errors during replication that prevent the molecule from copying itself correctly. In other words, it self-destructs.

However, Eigen and Schuster suggested that a solution to this problem existed in the form of the *hypercycle*. A hypercycle is a cyclic process in which a series of molecular species each act as catalysts for the next species in the sequence. Catalysts are chemicals that help a chemical reaction to proceed. So, in a hypercycle, chemical species A might help species B to form, species B might help species C to form, and C in turn might help A to form. The point is that each chemical species in this process might be relatively simple, so that error rates are not a problem. However, by adding extra steps in the cycle, the entire information content of the process can grow ever more complex.

It didn't take long before a number of researchers pointed out a flaw in the hypercycle theory. This is that it can easily be killed by 'parasitic' species. Suppose that at some point in the cycle one of the species catalyses not only the next member of the cycle but also another (parasitic) species. Let's call this parasitic species D. Then on each repeat of the hypercycle, the population of species D will increase at the expense of the entire cycle. Thus D acts as a dead end that eventually traps all the material involved.

A solution to this dilemma was discovered by the Dutch team of Boerlijst and Hogeweg. They showed that the spatial distribution of the chemicals is crucial. Instead of the entire 'world' cycling at once, the hypercycles became centred on various foci and produced outward spirals. Parasitic species tended to get pushed towards the outside edges of the spirals where they were removed by other processes.

What *is* life?

Scientists and philosophers have struggled with this question for generations. Among the many criteria that have been identified are:

- self-reproduction
- ability to interact with the environment
- ability to interact in non-trivial ways with other organisms

There is still a large body of opinion arguing that life on Earth is unique in the universe. To find another example, even microbes on Mars would establish whether we really are unique. The more science tackles the problem the clearer it becomes that life itself is a natural result of processes that are widespread in the universe.

People tend to assume that the forms of life we see around us are all that is possible. Without other examples to study it is difficult to know just how great the possible variations might be. In earlier chapters we saw that where direct experiments are impossible scientists are turning increasingly to simulation as a way forward.

Simulation is one of the key research methods of the new field of science called artificial life (see Chapter 3). The basic aim of alife research is to answer the question posed in the heading of this section. That is, its goal is to understand life and living processes in general. As the name implies, one test of our understanding would be to create artificial life forms. Many of the topics in this book really fall into this field of alife, which is also helping us to understand the origins of life here on earth.

Silicon Life Forms

Does life need to be carbon-based? As we saw earlier, carbon forms an astonishingly rich variety of complex molecules. Of these there is at least one, DNA, which can produce a detailed genetic code. It is not obvious that other elements can provide such a molecular code, which confines life to essentially the temperature zone of tepid water.

However, the fledgling discipline of artificial life suggests that new life forms may be created in computer environments. With IBM's Deep Blue currently the equivalent of world chess champion, the power of computers rivals that of the human brain in some domains.

Thomas Ray, in the Tierra computer program, introduced digital life forms to computer memory. These are little more than fragments of computer memory with properties of breeding, mutation and so on. While Tierra is at a very primitive stage of evolution, the drive to use evolutionary and learning systems in robotics will surely see more and more complex 'silicon' life forms.

The thing that is missing compared with real life is reproduction. We can imagine building machines that can take over the process of building machines—and perhaps ultimately leave us out of the picture. But we are still a long way from DNA, with its hundreds of megabits of genetic information in a *single molecule*.

What is consciousness?

At school we are taught about various definitions and characteristics of life. They usually include reproduction, something that our robots are a long way from doing. But most of us can probably think of other properties as being important in defining higher life forms. Of these, consciousness is the most intractable and puzzling.

Is consciousness a property of brains amenable to biological/physical explanation or is it something else, not

197

accessible to scientific enquiry? At the moment we can't say, and no accurate philosophical discussion is ever short. Let's look at some specific issues and see how they impinge on the likelihood of artificial consciousness.

An interesting approach is to imagine being able to build a mechanical nerve cell: it has the same input/output properties as a biological cell, but it's powered by a battery and uses connecting wires. We know from the work of many neuroscientists that we can insert wires or capillary tubes into cells and artificially activate them. Thus there does not seem to be any real difficulty in doing the swap. Now we keep adding artificial neurons, one by one, until our brain is entirely synthetic. Has consciousness disappeared? Extensive debate has not produced a clear answer, but readers might like to make up their own mind.

Even if we reject the notion that artificial implants will have no effect on consciousness, that there is some flashpoint beyond which consciousness vanishes, we still have to assume a very close link between the activity of the brain and consciousness: brain damage, fever, drugs, so many things that perturb brain chemistry perturb consciousness too. Even if we follow Ray Jackendoff and assume that there are separate mind–brain and mind–mind problems, we can't ignore the very tight coupling of conscious and neural behaviour.

It might be tempting to assume that some level of connectivity or size is the point at which a brain becomes conscious. Yet, as Jackendoff describes very lucidly, there is a lot of very complicated processing going on in a brain of which we are *not* conscious and never can become conscious—all the pre-processing that goes on in the sensory systems, for example. An interesting corollary here is sleep. The memories of dreams (admittedly faint and erratic) feel like conscious experiences, even though they were not.

Francis Crick, Nobel Laureate along with James Watson for discovery of the structure of DNA, has actually proposed that consciousness resides in a particular brain structure, also implicated in sleep control. But if it resides in a particular brain region, and isn't instead some global property, then perhaps that area could go wrong, get injured, leaving us a substantially functioning person but someone who is functioning more like an automaton than a person. It's hard to imagine what this might be like or even if it is possible.

Let's turn from the intractable problem of what consciousness *is* to what it is for. Again, we're no further down the track. There is evidence for it appearing in stages as we go up the animal kingdom, though. Birds will attack their reflection in a mirror or a pane of glass with no notion that this might not be another bird. A cat is more sanguine, quickly realising that its reflection is *not* another cat. We accept this since a cat will go into attack mode against a cat on the other side of a window where smell and, to a significant extent, hearing are no longer cues. But a cat is not self-aware It cannot make deductions about itself from its reflection. Chimpanzees, on the other hand, *are* self-aware: they realise from the reflection in a mirror that a mark put on their head while they are under anaesthetic is theirs. For our part, human language takes us a further, massive, step in understanding ourselves and our relationship with the world.

With the enigma of consciousness, we come to the end of our book. It has been a fascinating journey through a landscape in which we can see the broad outlines, even grasp the richness of the scenery. But, like our first trip by car through some exciting place, we yearn to go back on foot and see it close up. So it is with complexity. It will occupy the world's finest minds for many years to come.

NOTES

1. Tom Wolfe describes this incident in detail in his book *The Right Stuff*.
2. The Analytic Engine was the first (unsuccessful) attempt to build a programmable computer. It foundered largely because early nineteenth-century technology was not capable of the precision machining required. Ada Lovelace was Lord Byron's niece. Her achievements are commemorated in the programming language Ada, which is widely used in the United States and elsewhere for military grade software.
3. As an aside, temperature is most meaningful for gases in thermal equilibrium where we get a smooth variation in the energy of individual molecules with just one peak in the number of molecules at each energy. Lasers are far from equilibrium, with a very asymmetric distribution of energies.
4. There is just a shade of doubt here. As we discuss later in the book, computation is essentially reversible, consuming no energy. Without an energy constraint, we don't know how fast we can ultimately make a computer. But for computation to be reversible we have to save all the intermediate steps. Since we physically need matter of some kind, even if we get storage down to the atomic level with so-called nano-technology, we are still defeated by the exponential growth in computing demands. There is speculation that quantum computation may allow easier storage of intervening steps, but this is far from established at the time of writing.
5. We have glossed over some of the mathematical details here. The number of steps or the time on any computer will depend on the particular method or algorithm used to solve the problem. Here we are using the simplest 'brute-force' algorithms to illustrate the point.

6. If the number of pixels is N and the number of levels each pixel can occupy is L, then the number of pictures, P, is L^N

7. Quantum mechanics is somewhat different, introducing additional constraints on the accuracy of measurement.

8. From D.H. Meadows, D.L. Meadows, J. Randers and W.W. Behrens III (1972), *The Limits to Growth*, Universe, Washington.

9. The naming of the different modules, usually referred to as *areas*, is not straightforward. Names given by the early anatomists have stuck and other names have been grafted on. Suffice it to say that it is somewhat deeper into the brain than the areas where most of the visual input from the eye arrives.

10. This idea underlies a major compression algorithm for images and sound, predictive coding. Instead of transmitting everything in a signal, transmit only the differences against what you would predict.

11. See J. Poskanzer (1991), *Xantfarms Simple Ant Farm for XII*. The Internet address is http://www.acme.com/software/-xantfarm/.

12. The Coriolis force acts on objects moving in a rotating system, causing their direction of motion to curve. It is commonly assumed to be the reason that water rotates in a particular direction as it goes down the plughole in a given hemisphere.

13. Creationists and opponents of evolution have often used the evolution of the eye as an example of the 'impossibility' of evolution. One of the most exciting developments of recent times has been an awareness of just how powerful evolutionary techniques really are. In fact, a paper in the proceedings of the Royal Society of London a few years ago shows how it *is* possible to evolve the lens as an eye in a quite small number of generations given some plausible assumptions.

14. Wintermute was a large and somewhat ruthless artificial intelligence in *Neuromancer*.

15. Clever Hans was a horse who could do arithmetic! He would tap a front hoof the exact number of times equal to the product of two numbers. In fact what he was doing was just tapping until he detected some cues from his trainer's demeanour that he was supposed to stop.

16. Australia's Commonwealth Scientific and Industrial Research Organisation.

BIBLIOGRAPHY

Barnsley, M. (1988), *Fractal Cities*, Academic Press.

Batty, M. and Longley, P. (1994), *Fractal Cities*, Academic Press.

Bossomaier, T.R.J. and Green, D.G. (1997), *Complex Systems*, Cambridge University Press.

Casti, J. (1994), *Complexification*, HarperCollins.

Eldredge, N. (1995), *Reinventing Darwin*, Weidenfeld & Nicolson.

Hall, N.(199), *The New Scientist Guide to Chaos*, Penguin Books.

Hodges, A. (1997), 'Turing: Alan Turing, a natural philosopher', No. 3, *The Great Philosophers*, Weidenfeld & Nicolson.

Holland, J.H. (1992), *Adaptation in Natural and Artificial Systems*, MIT Press.

Kauffman, S. (1993), *The Origins of Order*, Oxford University Press.

Kelso, J.A.S. (1995), *Dynamic Patterns*, MIT Press.

Levy, S. (1992), *Artificial Life*, Penguin.

Mandelbrot, B.B. (1982), *The Fractal Geometry of Nature*, W.H. Freeman.

Maynard Smith, J. and Szathmary, E. (1995), *The Major Transitions in Evolution*, W.H. Freeman.

Mossel, B. (1973), *Where No Road Goes*, Enterprise Publications.

Penrose, R. (1995), *Shadows of the Mind*, Vintage.

Peitgen, H-O. and Saupe, D. (1988), *The Science of Fractal Images*, Springer Verlag.

Pinker, S. (1994), *The Language Instinct*, Allen Lane.

INDEX

203

GARDENING
—— WITH ——
BULBS

A Practical and
Inspirational Guide

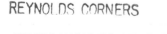

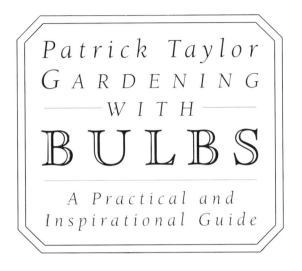

Patrick Taylor
GARDENING
—WITH—
BULBS

A Practical and
Inspirational Guide

Timber
Press

First published in 1996 by
PAVILION BOOKS LTD
26 Upper Ground, London SE1 9PD

Text copyright © 1996 Patrick Taylor

Illustrations copyright © 1996 Open Books Publishing Ltd

This book was devised and produced by
Open Books Publishing Ltd, Beaumont House
Wells BA5 2LD, Somerset, UK

Designer: Andrew Barron,
Andrew Barron and Collis Clements Associates

Computer Consultant: Mike Mepham

A CIP catalogue record for this book is available from the British
Library

Printed and Bound in Hong Kong

First published in North America by
Timber Press Inc
133 S.W. Second Avenue, Suite 450,
Portland, Oregon 97204-3527, U.S.A.
1-800-327-5680 (U.S.A. and Canada only)

ISBN: 0-88192-351-6

CONTENTS

❧

ACKNOWLEDGEMENTS

I am very grateful to many garden owners who allowed me to study and photograph bulbs in their gardens. I have received an immense amount of friendly help from one of the best bulb nurseries in Britain – Avon Bulbs, where Chris Ireland-Jones and Alan Street received my frequent invasions with the greatest friendliness and shared their deep knowledge most generously. I am very grateful to them. The Royal Botanic Gardens at Kew and at Edinburgh generously allowed me to photograph and gave me much help. Here I should like to thank Jenny Evans at Kew and Alan Bennell, John Main and Ron McBeath at Edinburgh. I learned much about the use of bulbs from the following gardens whose owners gave me generous help: Apple Court (Diana Grenfell and Roger Grounds), Ashtree Cottage (Wendy Lauderdale), Bosvigo House (Wendy and Michael Perry), Chenies Manor (Mrs MacLeod Matthews), Eastgrove Cottage Garden (Malcolm and Carol Skinner), Great Dixter (Christopher Lloyd), Greencombe (Miss Joan Loraine), Kingsdon (Patricia Marrow), Knightshayes Court (Michael Hickson), The Mead Nursery (Steve and Emma Lewis-Dale), The Monocot Nursery (Mike Salmon) and Wollerton Old Hall (Lesley Jenkins).

Patrick Taylor
Wells, Somerset

INTRODUCTION

☙

Bulbs are the most versatile of garden plants.
They will flourish in many different sites and
climates, and ornament any season. They are too
easily associated with spring – every day of the
year may be enlivened by their flowers. Apart
from the beauty of flowers many also possess
strikingly attractive foliage, whether it is the
exquisite marbling of *Cyclamen hederifolium* or
the bold architectural leaves of the larger irises.
Many are undemanding and will form
self-perpetuating colonies, happily co-existing
with other garden plants. They exist in every
imaginable colour and several are exquisitely
scented. In size they range from *Crocus minimus*,
scarcely 2in/5cm high, to the stately giant
Cardiocrinum giganteum up to 12ft/3.6m high.
Although some have very specific locations in the
wild, with climates of great extremes, they will
often prove remarkably accommodating, growing
vigorously in the very different artificial
environment of the garden.

Although in general I have concentrated on
bulbs that will flourish in most gardens and are
tough enough to survive the hurly-burly of the
border, very occasionally I describe plants that
may need special protection. Anyone who has
been to a well maintained Alpine house will know
the beauties of the diminutive species narcissus,
tulips, irises and fritillaries. Many of these have
such specific needs in the wild that they demand a
controlled environment when transplanted to a
foreign climate. Some of them, too, are so small
that they can easily be damaged when cultivating

the garden. However, some of the smaller bulbs, especially those having special cultivation needs, are marvellous to collect and grow in containers in the smaller garden. The miniature species crocuses, fritillaries, irises and tulips whose delicate charms may be swamped in the mixed border are wonderful plants to grow in this way.

This book describes a wide range of bulbous plants, with information on their cultivation, particular virtues and use in the garden. In the term 'bulbous plant' I include different categories of plant which the botanist defines much more precisely – bulbs, corms, rhizomes and tubers. They all have in common the ability to accumulate food and water to be consumed at an appropriate stage in the plant's life.

A bulb is a form of the base of a plant's leaves which has become swollen to provide a means of storing water and food. This allows it to become dormant with enough reserves to enable it to spring to life when seasonal conditions permit. If you cut across a bulb – an onion, for example – you will see the concentric layers of leaf of which it is composed. The fine, dry outermost layer of leaves protects the bulb from pests and infection. Some bulbs (daffodils, for example) are perennial, producing growth from the same bulb, repeatedly over many years. Others (tulips, for example) are annual, with flowers and foliage arising each year from a new bulb formed the previous year. Tulip bulbs grow at different rates and the size of the flower will be related to the size of the bulb. Bulb nurseries will grade bulbs so that a given batch will produce flowers of identical size.

Corms are underground, fleshy stems, resembling a flattened sphere, which provide enough sustenance for one year's growth only. When a flower has been produced a new corm will form above the old one. In addition baby corms – or cormels – are formed about the base of the old corm and perpetuate the plant. These, however, will take some time to grow into corms of flowering size. In appropriate conditions each

Snowdrops, Lenten hellebores and the decorative foliage of *Arum italicum* ssp. *italicum* 'Marmoratum' make a beautiful winter group

corm may produce many cormlets quickly building up a colony of plants. Colchicums, crocosmias, crocus and gladiolus are examples of corms.

Rhizomes are swollen stems, underground or half-submerged, usually horizontal in shape. They are perennial and spread by forming roots and throwing up flowering shoots from time to time. Many irises, some bamboos and Solomon's seal (*Polygonatum*) are examples of rhizomes.

A tuber is a swollen root or stem which may be either completely submerged or close to the surface of the ground. The potato is a characteristic root tuber forming new growth from the 'eyes' in its surface. Dahlias, some species of corydalis, and cyclamen are tubers.

Bulbs in the Garden

Bulbous plants ornament the garden throughout the year. To give something of the richness of plants available I shall go through the seasons briefly describing the bulbous highlights.

Even in deep winter bulbous plants will produce some of the most strikingly attractive foliage in the garden. December and January are brilliantly decorated by the gleaming marbled leaves of *Arum italicum* ssp. *italicum* 'Marmoratum' and the intricately shaped foliage of *Cyclamen hederifolium* whose surface is beautifully patterned in silver, grey and green. This foliage is not only decorative in its own right but provides a marvellous background for the explosion of spring flowers that is to come. In late January they are joined by snowdrops and aconites (*Eranthis hyemalis*) to form a marvellous winter picture. All these plants enjoy similar shady woodland conditions, the same conditions preferred by many herbaceous plants such as hellebores and pulmonarias which associate beautifully with them. Most gardens have areas of shade which are difficult to plant effectively. In quite small gardens it is perfectly possible to create a woodland corner, with plants such as these. They go so well together, and give an air of natural character, that it is probably best to create a naturalistic area for them rather than include them in the more formal parts of the garden where they can seem ill at ease. In large mixed borders, however, there will be places that may provide precisely the right growing conditions for bulbs. Deciduous shrubs, hydrangeas or magnolias, for example, will be leafless at the flowering time of snowdrops, winter aconites and *Cyclamen coum*. In the summer, when such bulbs are dormant, the dry shade under the shrubs' canopy is exactly what they need.

In March the explosion of daffodils gives never failing pleasure. Yet, however essential, they are difficult plants to fit in with other garden schemes. After their flowering the foliage must be left to wither, an unattractive sight in the border and not always easy to conceal. Some of the species, many of which are deliciously scented, make lovely plants for pots. But the hybrids, several of which are easy to naturalise, are really at their best in the orchard or other informal areas of long grass.

Early in the season bulbs give colour and attractive texture of foliage but it is only later that they begin to fulfill one of their other virtues – to provide structure. As border plants this is one of their most precious qualities. Crown imperials (*Fritillaria imperialis*), with

Snake's head fritillaries and blue and white bluebells flourish in the long grass of this orchard

their dramatically beautiful yellow or red-brown flowers in late March also have bold architectural presence. In my garden I grow them under the small tree *Amelanchier canadensis* which is just coming into leaf when they are in flower. They are underplanted with *Anemone blanda*, *Crocus tommasinianus* and *Cyclamen hederifolium*. In the summer they are followed by *Alstroemeria* Ligtu Hybrids, border phlox and *Tricyrtis formosana* all of which flower well under the light canopy of the amelanchier. Crown imperials are also excellent in a little formal box-edged parterre or as an early-flowering ingredient in the mixed border. Other fritillaries, in particular *F. persica* with its tall spires of almost black plum-coloured flowers and *F. raddeana*, not unlike the Crown Imperial but with creamy flowers, are also bulbs with strong structural presence. Other fritillaries have the chief quality of providing splashes of colour and often intricately shaped flowers. The snake's head fritillary, *F. meleagris*, with purple or white nodding flowers, is at its best in a

Illustration opposite:
Celandines (*Ranunculus
ficaria*) and the spring
squill (*Scilla vernum*)
growing wild in an English
wood

naturalistic setting. Flowering in April it is beautiful in the long grass of an orchard mixed with bluebells (*Hyacinthoides non-scripta*) which flower at the same time.

Overlapping with the fritillaries but continuing later are the erythroniums and trilliums. Many of the species of these two groups of plants relish the same woodland conditions. The foliage of both is among the most decorative of all – the bold marbling of trilliums and the curious watery markings of erythroniums. These exotic plants look wonderfully at home in woodland far from their native habitats. Most prefer rich, moist soil, looking marvellous with a background of moss and ferns. In the wrong conditions they will not merely look wrong but will also prove very difficult to cultivate. In my garden I grow erythroniums in a little bed against a north wall among the foliage of hellebores where, later in the season, hostas and the bold leaves of *Kirengeshoma palmata* take over.

The month of April shows the world of bulbs at its peak. Indeed, at this time of year it would be possible to make a beautiful garden using only bulbous plants. A walk in an English wood in this season shows the possibilities. Wild garlic (*Allium ursinum*) and bluebells (*Hyacinthoides non-scripta*), often growing together, form savoury-scented waves of blue and white under deciduous trees. The celandine, *Ranunculus ficaria*, is perhaps a little too wild for the domestic scale of the garden. On the edge of woodland it is often seen flourishing with the spring squill (*Scilla vernum*) making a sparkling mixture of gleaming gold and purple. In the more manicured parts of the garden the various cultivars of the celandine seem more appropriate, with their exotic flowers or foliage.

The various squills flower over a long period. I always relish the appearance of *Scilla mischtschenkoana*, bursting through the soil in February, its flowers starting to open as soon as they emerge. Both *S. messeniaca* and *S. liliohyacinthus* make

The wood anemone (*A. nemorosa*) with ferns in a woodland garden

marvellous underplanting in March and April. In the same season the brilliant violet-blue of *S. siberica* is lovely scattered in odd corners of the garden. Lastly, in June come the majestic plump drumsticks of *S. peruviana*, the only squill to have bold structural presence.

The anemones also make their contribution in March and April. I encourage blue or white *Anemone blanda* to wander at will in borders. They are not invasive, provide ornamental flowers and foliage, and happily share the ground with many other plants. They particularly flourish under a *Magnolia stellata* in a position which later in the season is taken over by daylilies and the giant *Lilium pardalinum*. But in April the mixed border is scarcely performing and this essential anemone makes an exquisite ornament. Its cousin the wood anemone, *A. nemorosa*, is better in a

The tulip 'Angélique' underplanted with variegated hosta

wild setting – among ferns at the edge of woodland, for example. It has very handsome foliage and will form a burgeoning mound, scattered with its crisp white flowers. There are pretty cultivars of the wood anemone (including the lovely violet-flowered *A. nemorosa* 'Allenii') but to my eye the best of all is 'Large-flowered Form' with its bold gleaming white flowers.

April sees the start of the tulip season. It is hard to imagine any gardener not including tulips in a list of essential garden plants. They are very versatile, available in a huge range of colours and span a wide flowering season – as long as three months. Some gardeners treat them as annuals, buying carefully graded new bulbs each year to produce reliable flowers of a uniform size. This labour-intensive and expensive form of gardening, choosing association plants with care, can produce lovely results. Tulips are admirable plants for containers in which it is possible to give them perfect growing conditions. In the mixed border they

Illustration opposite:
Kniphofia 'Sunningdale
Yellow' has a telling
effect in this exuberant
border of hollyhocks,
Rosa glauca, achilleas
and penstemons

can produce exactly the note of colour needed, associating well with the fresh new foliage of herbaceous and woody plants. If you can provide the right conditions naturalised tulips, in a wild setting, look magnificent. The dazzling yellow *T. sylvestris* or the scarlet late-flowering *T. sprengeri* are among the most beautiful tulips that have been successfully naturalised. Once having seen these flowering profusely in a meadow or at the edge of woodland it is easy to become rather impatient with fiddly arrangements in borders.

Irises flower over a long season and many of them, unlike tulips, also have the precious virtue of boldly architectural foliage. Some border irises have crisp fanshaped leaves which give valuable shape to a mixed border. Irises vary in character from the exquisite early flowering miniature alpine sorts to those like *I. pseudacorus* and *I. orientalis* which have magnificent blade-like foliage of dramatic architectural form. They vary, too, in their habitats, from moisture loving species like *I. sanguinea* to those like *I. unguicularis* which are at home in the driest of positions. This last, producing its sweetly scented flowers in December, and for weeks afterwards, must be near the top of every gardener's list of essential plants.

Overlapping with the early tulip season are the camassias. They are not difficult plants but are surprisingly underused in gardens. They are all good in the border and one, *C. quamash*, is excellent for the rough grass of meadows or orchards. The double creamy-white *C. leichtlinii* 'Semiplena', flowering in May, is among the most beautiful plants in its season. Other members of the Liliaceae family also make their mark in May or June. The rather tender *Asphodelus aestivus* is worth taking a lot of trouble for. It sends forth in June bold branching spires of white flowers – a superlative, shapely border plant which associates effortlessly with almost anything else. The anthericums and paradiseas are other liliaceous plants which also produce strikingly shaped decorative white spires of

flowers in this season.

In June, when many gardens are thought to be at their peak, and a vast range of non-bulbous plants has come into full flower, bulbs still have a contribution to make. Some of these are as bold in structural emphasis as they are in colour. The foxtail lilies (*Eremurus* species) throw out soaring spires of flowers, rising as high as 8ft/2.5m to make marvellous aerial ornaments. The African pokers – *Kniphofia* species – also make a valuable dual contribution of colour and form. Some of these will flower over a very long period, starting in May and continuing throughout the summer. Some are very tall but even the small ones have well-defined presence in a busy border. Very late in the season, in August and September, the larger kniphofias are among the most magnificent border plants. Fashionable garden taste rather turned against them in recent years but sense has prevailed and their brilliant qualities are properly appreciated once again.

Lilies are a particularly precious group of plants, strikingly at home in very different contexts. Some, like *Lilium martagon*, are best in a wild setting: it is one of the best lilies for naturalising in dappled woodland shade. Others, even other species like the magnificent *L. pardalinum*, are superb border plants. Lilies are excellent for the richly planted scheme where their bold and brilliant flowers make an emphatic statement in the profusion of foliage and colour. *L. regale* will harmonise with many arrangements and possesses one of the finest scents of any flower. Lilies, too, perform very well in pots and some of the tender, sweetly scented ones (such as *L. longiflorum*) will ornament a sunny terrace to wonderful effect. Lastly, despite their reputation for naturalising only with difficulty, the species are remarkably easy to propagate from seed, producing flowering bulbs quite quickly.

Crocosmias, formerly known as montbretias, are also plants of high summer which continue to make a contribution over a long period. New cultivars are constantly appearing and some, like the magnificent

A spectacular late-summer border at Great Dixter with blood-red dahlias, the bold leaves of cannas and purple *Verbena bonariensis*

'Lucifer', are plants of tremendous character. They all, of whatever size, have valuable sword-shaped foliage which can give valuable structural emphasis as the mixed border becomes rather formless in late summer. Nor are they all orange and scarlet; some of the yellow forms, such as the late-flowering 'Golden Fleece', are beautiful.

Some of the plants which I describe later in the book are too tender for all but the mildest gardens. But in less favoured climates they make marvellous plants for pots which gives them the additional attraction of providing mobile ornaments. The fabulous African *Albuca nelsonii*, with statuesque foliage and vanilla-scented white flowers, is magnificent in a pot. The eucomis tribe, also from Africa, has similar virtues but, alas, most species smell horrible. But they are beautiful and they may be bedded out in the border very satisfactorily. Some of the more tender gladiolus – like the sweetly scented *Gladiolus callianthus* – may be planted in quantity in a large pot where it will form a sheaf of exquisite white flowers. If well looked after, the corms will multiply with surprising ease. The mysterious almost black *Cosmos atrosanguineus*, with chocolate scented flowers, will in most gardens flourish far more vigorously in a pot than planted out. All these will be flowering in high summer to make wonderful ornaments for the terrace.

Late summer is almost as rich in bulbous plants as early spring. Although some of these (such as the stately galtonias) are of subdued colouring many more explode in dramatic reds, purples and yellows. Garden taste is once more turning to these exciting arrangements to which bulbous plants make such a contribution. Dahlias of the deepest red (such as *D.* 'Arabian Night'), the giant foliage and flamboyant flowers of cannas, and the sharp raspberry red of the little *Alstroemeria psittacina* are dazzling ingredients in the 'hot' border.

With autumn the bulbous season is by no means finished. The garden now sparkles with enchanting nerines, *Cyclamen hederifolium* scattered like jewels in unlikely places, colchicums and sternbergias producing exquisite flowers among the first fallen leaves, and flamboyant amaryllis throwing out their candy pink flowers. As the first frosts clear away tender growth, the exquisite foliage of cyclamens is revealed and in December the indomitable flowers of *Iris unguicularis* appear to remind us that the great bulbous cycle is about to start again.

This brisk tour through the year has only touched upon the major groups and a few others. But there are many more which I have not mentioned, and these are described in detail in the directory of plants in the following pages. Hardiness zone ratings are given for each plant and maps on pages 248–249 show which geographical areas fall into which zones. The hardiness rating, however, is only a broad indication and gardeners should be encouraged to experiment in their own garden. Until you have tried it yourself you can never be certain how a plant will respond.

Albuca

There are about 30 species of albuca, all bulbs, in the family Hyacinthaceae/Liliaceae, most of which are native to South Africa.

Albuca nelsonii
Origin: South Africa
Z: 10

❧ This lovely plant is probably too tender for any but the most privileged garden. However, I have heard of gardeners getting away with it in a well-drained sunny position in Zone 9 with a deep winter mulch. In less favoured gardens it makes a superlative pot plant for the cool greenhouse where it will survive occasional temperatures as low as −5°. It produces a striking plume of glistening green leaves which curve outwards from the centre. A fleshy flowering stem, up to 36in/90cm tall, erupts from the middle bearing a raceme of flowers. The buds are striped with green and the flowers are a waxy white when they open, giving off an exotic scent of vanilla and liquorice. In growth it should be lavishly fed and kept dry when it becomes dormant in late summer or early autumn. On a terrace, especially near a sunny sitting place where its delicious scent may be savoured, there are few more exquisite bulbous plants. Its statuesque shape makes it especially suitable to plant in pots to flank an entrance. In hot weather in the full sun they will need to be watered at least daily if they are in terra-cotta pots.

Allium

The onion tribe is a vast one – with at least 700 species of bulbs and rhizomes in the family Alliaceae/Liliaceae very widely distributed but found only in the Northern Hemisphere. Apart from producing some of the essential culinary flavourings such as the true onion (*A. cepa*), the leek (*A. ampeloprasum* var. *porrum*) and garlic (*A. sativum*), many are also valuable decorative plants. Some, such as chives (*A. schoenoprasum*), are delicious herbs that also produce ornamental flowers – not showy enough, perhaps, for the flower garden but a welcome bonus in

the herb or kitchen garden. Some are far too invasive
for all but the wildest garden; the pretty rosy garlic (*A.
roseum*) runs amok in my garden – apparently
reproducing by every means known to botanists.

Allium acuminatum
Origin: N.W. America
Z: 6

❧ For reasons unknown to me this exceptionally
pretty onion has never caught on as a fashionable
garden plant. Its flowers have dashing character and, in
relation to the umbel into which they are gathered, are
larger than usual, giving the plant a lively appearance.
Opening in June they are a rich deep rosy pink with
graceful pointed petals with a deeper stripe down the
centre. Up to 20 flowers are gathered in the umbel
which, because of the size of the flowers, lacks the crisp
drumstick outline of many alliums. The flowering stem
rises no more than 12in/30cm, with slender leaves
gathered about its base. It does best in rich soil in sun
or part-shade. There is a white clone which lacks the
charm of the more usual pink form. It may be
propagated by seed or by division of bulbs in the
autumn.

Allium aflatunense
Origin: Central Asia
Z: 7

❧ The various ornamental onions which form flowers
like emphatic drumsticks are immensely valuable
garden plants. *A. aflatunense* has handsome glaucous
leaves from which erupts a fleshy stem crowned in May

Illustration: *Allium aflatunense* 'Purple Sensation'

by 4in/10cm spheres of purple flowers. The spheres are composed of an immense number of diminutive flowers each held at the tip of a rose-coloured stem. Purple anthers protrude far beyond the tips of the petals. The type is a pretty enough plant but a cultivar with a very off-putting name, *A. aflatunense.* 'Purple Sensation' is a tremendous improvement. Not only is it much taller, up to 36in/90cm, but the flowers are a rich and vivid violet – the sort of colour a rather racy cardinal might wear. It is magnificent in a colour scheme of purples and reds and looks especially beautiful rising from the fresh glaucous foliage of certain hostas (such as *Hosta* 'Krossa Regal'). It will do best in rich, moist soil and will flourish in the semi-shade.

Allium caeruleum
Origin: Asia
Z: 7

❧ This rich blue allium is a native of the steppes. Its stems are slender, rising as high as 24in/60cm, with in June a spherical umbel of flowers, 1 1/2in/4cm in diameter. Each flower is deep blue with an even deeper coloured vein, and blue anthers. By the time it flowers its insignificant grass-like leaves will have withered. It must have a sunny, dry position. It looks very beautiful among smaller shrubs that also have a touch of blue in their colouring such as *Artemisia* 'Powis Castle' or with sea-hollies like the steely blue *Eryngium × zabelii*. It is easily propagated by seed. A form of it, *A. caeruleum* var. *bulbiferum*, also produces bulbils on the flower head.

Allium cernuum
Origin: N. America
Z: 6

❧ This little allium, very widely distributed between Canada and Mexico, has a graceful shape of flower unlike any other. The flowers are arranged in a loose umbel carried at the tip of a stem which curves over at the top. Each flower hangs at the end of a stem which curves upwards and then plunges downwards, so that the flowers resemble little bunches of exotic fruit. The flowers are shaped like rounded lanterns, with white

Illustration opposite:
Allium christophii

stamens protruding. They vary in colour from a rich pink to a deep maroon with hints of magenta. The flowers stems, which are coloured with a plummy bloom, are very slender, allowing the flowers to sway in the slightest breeze. The seed heads are strikingly decorative, with the flower stems stiffening and the seed-pods appearing to shoot outwards like an exploding firework. The flower stems will rise to about 18in/45cm. There is a white cultivar, *A. cernuum* 'Album', not so exciting as the type, and an excellent rich violet one, *A. cernuum* 'Hidcote', which is worth seeking out.

Allium christophii
Origin: Central Asia, Iran, Turkey
Z: 7

❧ This, formerly known as *A. albopilosum*, is one of the most handsomely decorative of the alliums. Its spherical flower-heads, up to 9in/23cm in diameter, are carried on thick fleshy stems up to 24in/60cm tall. The flower buds are lime-green and pointed, opening out gradually in June to form star-shaped flowers with very fine petals. They are violet in colour with a striking metallic sheen, and purple stamens stick straight out from the centre of the flowers. The plumpness of the flower-head is refined by the airy delicacy of the flowers that cover it. The leaves are rather lax, wide, strap-like and glistening. Its colour enables it to mix with many schemes. I have it growing most ornamentally among the brick-red flowers of *Euphorbia griffithii* and it is marvellous among geraniums. After flowering, the seed heads are ornamental. It should have a position in full sun or half shade and it is best in rich soil. It will seed itself gently.

Allium cyathoporum var. *farreri*
Origin: China
Z: 8

❧ Formerly known as *A. farreri* this is among the smaller alliums, rising to no more than 8in/20cm, and is a pretty front-of-the-border plant. The flowers open in June, hanging tassels of sprightly purple flowers each of which has sharply pointed petals giving the flower heads a slightly prickly look. The foliage grows in thick clumps of shining, grass-like leaves. In heavy, wet soil it will seed and multiply, but not invasively. It may be propagated by dividing clumps. It is very decorative in

Illustration: *Allium cyathoporum* var. *farreri*

a sunny position at the front of a border, with pinks and the smaller geraniums. The flower colour is especially attractive against grey. The small artemisia, *A. alba* 'Canescens', which produces a fine froth of silver foliage, makes the perfect background.

Allium flavum
Origin: Asia, Europe
Z: 7

❧ Although this onion forms the drumstick umbel of flowers common to so many other species, the form of the individual flowers is so striking that it is transformed into a plant of very different character. The flowers, which appear in June, are golden yellow, star-shaped, with shapely pointed petals that have an elegant curve. There may be as many as 50 flowers on each umbel which is carried on a stem up to 12in/30cm tall. Each petal has a stripe of green down the centre. Some authorities claim that it is sweetly scented, all I can say is that the clone in my garden has no scent at all. The leaves are decorative, broad, undulating and glaucous green, resembling a tulip's. It will flower best in rich soil in a position in the sun or in part-shade where its glowing colour is seen to best advantage. It seeds itself or clumps may be divided in the autumn.

Allium karataviense
Origin: C. Asia
Z: 8

❧ There is nothing like this immensely decorative allium which is equally striking in foliage and flower. The leaves are very broad, up to 4in/10cm, curving, with a curious metallic sheen, grey flushed with purple.

Illlustration: *Allium flavum*

The spherical umbels of flowers are up to 8in/20cm in diameter, opening in May, composed of countless diminutive stars, a pale violet-grey in colour, becoming paler as they age. The umbels are carried on relatively short stems, no more than 12in/30cm high, giving them the appearance of nestling among the exotic leaves. This is one of the less hardy alliums but so beautiful that it is worth taking trouble to provide the right conditions for it. It must have a sunny position and will not tolerate heavy water-logged soil. It is beautiful growing among other small herbaceous plants, such as geraniums. It looks lovely among the soft glaucous grey leaves and pale violet flowers of *Geranium renardii* which flowers at the same time.

Allium moly
Origin: Europe
Z: 7

❧ This yellow onion has a character all of its own. Its little flowers, opening in June, are grouped in loose umbels but face upwards. Each flower is a resplendent rich yellow star with well separated petals and paler yellow stamens. As the flowers age they become a much paler colour. The plant rises up to 12in/30cm high and the foliage is strikingly handsome, with glaucous grey leaves as broad as 1 1/2in/4cm. In the wild it is found in poor soil in shady places, a position easy to match in most gardens. A shaded site has the additional advantage of preserving the lively yellow of the flowers. It needs sharp drainage. It may be propagated by seed or by dividing the bulbs in autumn. The handsome cultivar *A. moly* 'Jeannine' is a larger plant sometimes producing two generous umbels on a single stem.

Allium nigrum
Origin: Mediterranean
Z: 8

❧ This splendid Mediterranean allium is not often seen but it makes a marvellous garden plant. The flowering stems rise as high as 36in/90cm bearing at their tips stately umbels of flowers shaped either like a broad shuttlecock or a flattened sphere. The flowers, which open in June, are white and their petals curve forward slightly and are marked with a green stripe, and the anthers are a pale yellow, which gives the whole flowerhead a greenish-golden cast. The seeds are

striking – a glistening green or black – appearing as the flowers are still in place. The leaves, grouped together at the base of the flower stems, are very broad, up to 6in/15cm, a glistening fresh green. In the wild this is a plant found in poor soil in waste-land. It needs a sunny position and makes an admirable border plant with its mysterious flower heads rising among other plantings. It will seed itself gently or it may be propagated by dividing the bulbs.

Allium obliquum
Origin: E. Europe, Central Asia
Z: 7

No other allium has quite the colouring of this decorative plant. The flowers in June are spheres, 1 1/2in/4cm in diameter, of pale lime-green flowerlets, like tiny Chinese lanterns, with stamens protruding far beyond the tips of the petals which gives the whole flower head a hairy appearance. The colouring is in appearance like a cold glass of Chablis. The flowering stems, which rise as high as 24in/60cm, are blue-green as is the grass-like foliage. It is best in a sunny position in rich soil and looks very pretty growing among the sprawling foliage of pale creamy-yellow *Scabiosa*

Illustration opposite:
Allium rosenbachianum

ochroleuca or pale violet *Scabiosa columbaria*. This is
the kind of small plant of character which, without
being showy, makes a decorative contribution to odd
corners of the garden. It may be propagated by seed or
by division.

Allium rosenbachianum
Origin: Central Asia
Z: 8

🐌 This is one of the taller ornamental onions whose
flower-head sways elegantly at the tip of a slender stem
up to 36in/90cm high. The flowers are gathered
together into a spherical umbel up to 4in/10cm in
diameter. Opening in May, each flower is a lilac-purple
star with a metallic sheen. The rather prominent
stamens giving the flower-head a diffused outline. As
the flowers fade they are replaced by jade green seed
pods which contrast strikingly with newly opened
flowers. It makes a wonderful border plant combining
great presence with delicacy. It looks beautiful with
grey-leafed plants and I have seen it gleaming among
the very pale foliage of *Elaeagnus commutata*.

Allium roseum
Origin: Mediterranean
Z: 8

❧ I include the rosy garlic with some misgivings for it has found in our garden, with its heavy, moist soil, the perfect home where it reproduces with prodigal abandon. The leaves appear before the flowers, very thin, like chives. The flowering stems rise 18in/45cm high crowned by a loose umbel of flowers 3in/8cm across which open in May but continue for weeks. They are a fine delicate rose-pink, up to 30 gathered together each at the tip of a wiry stem. As the flowers fade bulbils are produced at the base of each umbel. They become deep red, glistening like something deliciously edible (I do not recommend that you try to eat them). These will be scattered and many new plants produced. The bulb underground also multiplies alarmingly. Although invasive, the foliage is inconspicuous and the flowers, which will intermingle with low-growing plants, are very decorative. In my garden a colony has established itself among very deep purple aquilegias which mix attractively with the pink flowers. It will grow almost anywhere.

Allium schoenoprasum
Origin: Asia, North
America, Europe
Z: 5

❧ The culinary chive is one of the most widely distributed of the allium tribe. Apart from the fact that it is one of the few truly essential herbs it also has great decorative potential, especially in small gardens. The flowers are produced in May or June, small spherical umbels of diminutive flowers of a soft blue-purple held at the tips of very slender, bright green stems which rise up to 9in/23cm. The flowering stems, by the way, are too woody for kitchen use. The finest flavoured shoots are those that first appear, or the secondary growth that will spring up after the first cutting – older growth is coarse in flavour. The decorative value of chives has long been recognised in old-fashioned kitchen gardens where they were often planted in neat rows along the edge of paths. It will grow in shade or in full sun but it should have a rich, moist soil. It is very easy to propagate by seed or by dividing clumps in the autumn. A white form, *A. schoenoprasum* 'Album', also known as 'Corsican White', is particularly pretty and the pink, *A. schoenoprasum roseum* also has its charms.

Allium schubertii
Origin: E. Mediterranean
to C. Asia
Z: 8

❧ This onion is a bizarre but beautiful oddity. The bulb nursery Avon Bulbs describes it as resembling 'a frozen firework in mid burst' and I cannot improve on that. It produces in June a large umbel of flowers, up to 18in/45cm in diameter, among the most imposing of any onion. The individual flowers are star-shaped, a

warm pink in colour but their oddity is that they are
borne on stems of varying length, some very short but
others up to 8in/20cm long. The flower stems are a
vivid jade green, making a lively contrast with the pink
of the flowers. When the flowers have faded the dead
seed-heads are strikingly decorative, with the stiff stems
shooting out varying distances from the centre. Flower
arrangers go mad about them. It needs a sunny position
and will not survive waterlogged soil, needing to dry
out completely in dormancy. Plant the bulbs a good
4in/10cm deep and in colder gardens a deep mulch will
help to protect them. It is easily propagated by seed
which is produced in quantity.

Allium sphaerocephalon
Origin: Europe
Z: 5

❧ The round-headed leek is a European native found
in rocky and grassy places, usually in limestone
country. It is among the later-flowering of the alliums,
producing its flowers in July and August. These are
spherical umbels of a vivid red-purple verging on
crimson. The top of the umbel is slightly pointed giving
it somewhat the shape of an onion and is held at the tip
of a stem up to 24in/60cm tall. The flower-head is
heavy in relation to the stem, causing it to sway freely
in a breeze. Bees love the flowers which can often be
seen springing up and down as the bees descend or take
off. It makes an excellent border plant, in full sun or
part-shade and it will grow in almost any soil. It will
enliven many schemes, hotter ones of reds and purples
as well as cooler arrangements of pinks and blues.
Wishy-washy violet flowers are occasionally seen,
much duller than the more common rich purple. There
is also a white cultivar. It may be propagated by seed or
by dividing the bulbs.

Allium triquetrum
Origin: S. Europe
Z: 6

❧ The three-cornered leek derives its name from the
sharply triangular section of its stems. It is one of the
most beautiful of the more invasive alliums – few plants
of such thuggish disposition have such elegant flowers.
These appear in April or May, umbels of sprightly
white trumpets, with backward-curving petals crisply
marked with green stripes. The umbels hang on one

side of the stem only, with as many as fifteen flowers.
The foliage is fresh green, forming bold sheaves among
which the flowering stems rise. Stems and leaves give
off a rich onion scent when bruised. In the wild it is
found in moist shady places in woodland or by the
banks of streams. In the garden it will flourish in the
shade making a marvellous companion for the fresh
new foliage of ferns, Asiatic primulas, *Brunnera
macrophylla* and other plants enjoying similar
conditions. It will spread quite vigorously and seed
itself gently.

Allium unifolium
Origin: California, Oregor
Z: 8

❧ Formerly known as *A. murrayanum* this beautiful
American north-west coast species is a very decorative
garden plant, though not the hardiest of its tribe. It
flowers in early June, producing a loosely-formed
umbel of rosy pink flowers. The flowers are marked
with a deeper stripe down the middle of the petal and
as they age they become much paler, producing an
attractive mottled effect of different shades of pink. It
grows to a height of about 18in/45cm and, despite its
name, certainly produces more than one leaf! It needs a
sunny, well-drained site. I have seen it well used at the

Illustration opposite:
Allium ursinum

Allium ursinum
Origin: Europe, Russia
Z: 5

front of a border rising behind clumps of cottage pinks.
It is beautiful growing through the pale silver-grey
foliage of *Artemisia* 'Powis Castle' which enjoys
exactly the same conditions.

❧ In cool woods in England, flowering in April at the
same time as bluebells (*Hyacinthoides non-scripta*), the
wild garlic, or rampons, fills the air with the bracing
whiff of garlic. It does not at all resemble true garlic
(*A. sativum*) in growth. Distinguished tufts of gleaming
leaves – broad, ribbed and undulating – form billowing
mounds of foliage. From their centre the flowers
emerge in April, spherical umbels at the tips of fleshy
stems which rise high above the foliage. The umbels are
composed of diminutive star-shaped flowers, white
with a green eye. Wild garlic, like the bluebells so
inevitably associated with it, is a plant of shady
woodland, relishing deep, moist soil. In the wild it is
also found in hedgerows and on shady banks. In the
garden it is emphatically not a plant for the border. In
an area of wild woodland character, flourishing in deep
shade, it has tremendous character to which is added
the rich smell of garlic, one of the most vivid scents of
spring. You do not need to know how to propagate it –
it will do it for you all too obligingly.

Alstroemeria

There are about 50 species of alstroemeria in the family Alstroemeriaceae/Liliaceae. They are all perennials with rhizomes or tubers.

Alstroemeria aurea
Origin: Chile
Z: 7

❧ This is one of the hardier alstroemerias, formerly known as *A. aurantiaca*. It will make a substantial clump, invariably needing support, rising to 3–4ft/90–120cm high. The leaves are glaucous-grey with an attractive undulating form. The flowers in July are a marvellous warm gold, shaped like an irregular trumpet, freckled inside with deep red spots. They continue well into August and make one of the best late-summer herbaceous plants for the border. I have seen it planted with the almost black red dahlia 'Arabian Night' which echoes the red of the alstroemeria's spots. It is equally valuable in cooler schemes of cream and pale yellow. It will flower well in sun or dappled shade but must have good drainage. It may be propagated by dividing the tubers. The cultivar *A. aurea* 'Dover Orange' is a larger plant with rich orange flowers.

Alstroemeria haemantha
Origin: Chile
Z: 9

❧ This rather tender alstroemeria is rare, but worth seeking out for it has flowers of the most beautiful colour. They open in June, loose trumpets in groups held at the tips of wiry stems. They are a vibrant red-orange in colour and have a presence out of proportion to their size. Flowering stems rise up to 24in/60cm and the foliage is an attractive glaucous grey. It must have a sunny site and requires humus-rich soil. In the garden it would make a superlative contribution to an arrangement of reds, oranges and yellows. I have seen it looking magnificent against the silver new foliage of *Artemisia* 'Powis Castle.' It may be propagated by dividing the tubers in autumn or by seed which may be much easier to find than the plant itself.

Alstroemeria Ligtu Hybrids
Origin: garden
Z: 8

❧ The origins of this garden plant are lost in the mists of hybridisation. It has splendid glaucous leaves, long, narrow and twisting. The fleshy stems, which rise as high as 5ft/1.5m with suitable support, bear lavish sprays of flowers in June. These are irregularly trumpet-shaped and vary in colour from creamy white, yellow to blushful pink. All have a single petal that is marked with bold stripes of a contrasting colour. The

result can seem like an explosion in a candy factory but
in the right setting the plant has lavish charm, with an
air of insouciant abundance. Its habit of growth is lax
and it really needs support which in a mixed border
could be provided by adjacent woody plants through
which it will weave decoratively. It will grow in sun or
part shade and needs a rich, moist soil. In full sun some
of the colouring will be lost. It may be propagated by
division in the autumn.

Alstroemeria psittacina
Origin: Brazil
Z: 8

🙢 Previously known as *A. pulchella* this is one of the
most garden-worthy of the alstroemerias. Its fleshy
tubers stir into life in May, thrusting out pale green
leaves followed by stiff, almost woody flowering stems,
up to 24in/60cm tall, bearing leaves at regular intervals.
The flowers, which open in July, are of splendid
distinction. They hang sideways, pointing slightly
upwards, long and loosely trumpet shaped (actually
more resembling a clumsily furled umbrella). They are
a fresh raspberry red in colour, up to 3in/8cm long, and

the petals are tipped with pistachio green. The petals are speckled within and without and the anthers are the palest green. Few bulbous plants provide such a beautiful display in late summer, continuing in flower well into August. I have seen it brilliantly used *en masse* in a 'hot' border with crocosmias, dahlias, orange daylilies and flaming red floribunda roses. But it will make an admirable front-of-the-border plant in many schemes. It needs a sunny position in well-drained soil. Slugs are a problem which may be solved by incorporating grit into the planting soil. In colder gardens a deep mulch will protect it in hard winters. It may be propagated by dividing the clumps of tubers in autumn or spring.

× **Amarcrinum**

× *Amarcrinum memoria-corsii*
Origin: Garden
Z: 8

This is a cross between the two genera, Amaryllis and Crinum, in the family Amaryllidaceae/ Liliaceae. The same cross has also been called × *Crinodonna.* There are two species.

ᘓᕐ This spectacular late-flowering bulb has great character. It flowers in August producing bold upward-pointing umbels composed of many trumpet-shaped flowers. The flowers are a lively shell-pink with petals that twist slightly and they give off a sweet scent. The flower stems are thick and fleshy, rising to a height of 30in/75cm, erupting from a sheaf of slender leaves. With its flowers pointing towards the sky, there is something triumphant about this plant. It needs the sunniest place in the garden, with protection from a wall. It will do well in quite poor soil but it absolutely demands sharp drainage. It will flower for several weeks and make one of the most striking ornamental plants in the last weeks of summer and the first of autumn. It makes a magnificent plant for a large pot.

Amaryllis

There is a single species of amaryllis, in the family Amaryllidaceae/Liliaceae. The name is often misused for the forced cultivars of *Hippeastrum* often seen for sale in shops at Christmas time.

Amaryllis bella-donna
Origin: South Africa
Z: 8

This splendidly glamorous bulb is a marvellous ornament in the autumn garden. It throws out fleshy, purple-tinged stems which rise up to 36in/90cm and the flowers appear in September or October. They are carried in groups of up to six handsome trumpets of a rosy-pink verging on purple, each 4in/10cm long. The petals are pointed and curve back to reveal a pale throat and long stamens. The flowers are marvellously scented and last for weeks, often remaining well into November. The strap-like leaves appear after the flowers and remains throughout the winter. They should not be removed until the following summer when they are completely browned. It must have the sunniest site you can find and in severe winters it is helpful to protect the foliage with a loose covering of bracken or something similar. The bulbs should be planted quite shallow, with the necks just level with the surface of the ground. In congested groups the bulbs are often forced slightly above ground which, as with crinums, seems to encourage flowering. It lends itself very well to cultivation in pots. It may be propagated by removing bulb offsets. There are several cultivars: *A. bella-donna* 'Hathor' is very floriferous with peachy-pink flowers; *A. bella-donna* 'Johannesburg' has rich pink flowers.

Anemone

There are 120 species of anemones, in the family Ranunculaceae, all of which are herbaceous perennials and those described below are rhizomatous.

Anemone blanda
Origin: S.E. Europe
Z: 5

❦ In its native habitat *Anemone blanda* is found in rocky places and scrubland in dry, mountainous regions, up to 6,000ft/2,000m. But it will flourish in very different conditions such as those in my own garden, which has heavy soil and high rainfall. The flowers are single, with up to fifteen petals – blue, white or pinkish-violet – appearing in March, unfolding from tubular flower-heads. They are variable in size – from 3/4–1 1/2in/2–5cm. The leaves are intricately lobed and form a mound above which the flowers rise. *Anemone blanda* will flourish in part shade or under deciduous trees and shrubs. It is also very attractive in short grass – I have seen it in an orchard under apple trees. But do not plant it where animals graze, for it contains a narcotic, anemonin, dangerous to livestock. There are some good cultivars including a handsome white one, *A. blanda* 'White Splendour', with larger flowers than the type. Cultivars should be propagated by division in the spring but the type, in an appropriate position, will seed itself univasively, producing attractive variations in colour. *Anemone appenina*, with clear blue flowers, is similar except that it has slightly hairy foliage.

Anemone × lipsiensis
Origin: Garden
Z: 4

❧ This hybrid between *A. nemorosa* and *A. ranunculoides* produces its flowers in April – single, 1in/2.5cm across, a beautiful pale cream-yellow with lemon-coloured anthers. The flowers are borne well above the leaves which are lobed and slender with a red tinge to their stems. I have seen it planted on the edge of a small woodland garden intermingled with white forms of the common primrose (*Primula vulgaris*), conveying the essence of spring. It is also good in a rock garden, in an elevated position so that its detail may be appreciated. It will come true from seed.

Illustration: *Anemone nemorosa* 'Large-flowered Form'

Anemone nemorosa
Origin: Europe
Z: 5

❧ The European wood anemone is a beautiful flower of hedgerows and woodland usually in limestone country. It has prettily divided leaves and single white flowers appearing in March about 1in/2.5cm across with lemon yellow stamens. Various cultivars provide the most attractive garden plants. A group of clones, known simply as 'Large-flowered Form' has strikingly bigger flowers, up to 2in/5cm in diameter, and makes an especially good plant for a woodland setting. The double-flowered *A. nemorosa* 'Alba Plena' is an ancient garden plant, known since at least the 17th century. It has a central ruff of petals surrounded by single petals. *A. nemorosa* 'Allenii' has single flowers larger than the type, a good lavender-blue. *A. nemorosa* 'Robinsoniana', named after the great gardener and writer William Robinson, has similar flowers, not quite so large, carried on striking red-brown shoots. All these will do best in rich, moisture-retentive soil.

Anemone ranunculoides
Origin: Europe
Z: 4

❧ The yellow wood anemone is found very widely in Europe – from Russia to Spain. It has especially attractive fresh green foliage, with each leaf decoratively lobed, like a slender oak leaf. The flowers in April are an astringent yellow, with rounded petals and a silken texture up to 3/4in/2cm across, with a pale green eye. The flower stems are covered in silver hair and the half-open bud, hanging gracefully downwards, is exquisitely formed. This is a plant of deciduous woodland and in the garden it is easy to find a good place for it. In a suitable position it will form a spreading clump which, with its fresh colouring, will illuminate any part of the garden. It may be propagated by division in the autumn. There is a double-flowered cultivar, *A. ranunculoides* 'Pleniflora'.

Anthericum

There are about 50 species of anthericum, all rhizomes, in the family Asphodelaceae/Liliaceae.

Anthericum liliago
Origin: Mediterranean
Z: 7

❧ The St Bernard's Lily is one of the most attractive spring-flowering bulbs, consorting easily with other plants. It is not unlike a white-flowered asphodel but with a character entirely of its own. The flower heads, in the form of racemes up to 10in/35cm long, resemble elongated heads of wheat. They are held on stems up to 24in/60cm tall. The flowers which open in May are single, white with a dazzling silken texture. The yellow anthers are very prominent, thrust forward at the tips of white stamens. The leaves are grey-green, rather lax, forming a busy mound above which the flowering stems rise high. In the wild it is found in high meadows, woodland or quite dry, stony places. It will flower well in part-shade. It is a beautiful partner for other late spring bulbs, in particular tulips. I have seen it planted with the tulip 'Spring Green' whose flowers are a subtle creamy white flushed with pale green.

Arum

There are about 25 species of arum, all tubers, in the family Araceae.

Arum italicum ssp.
italicum 'Marmoratum'
Origin: Europe
Z: 6

There are very few bulbous plants whose chief beauty is their foliage. This arum is among the most valuable, strikingly decorative yet flourishing in awkward corners, such as dry shade. The leaves appear in autumn and make a marvellous winter ornament. They are roughly arrow-shaped, up to 12in/30cm long and quite narrow, with an undulating surface which is a lustrous dark green intricately marked with a marbling of pale green veins which stop just before the edge of the leaf leaving a crisply delineated margin. The whole plant, at its most vigorous in early spring, makes a shapely mass of leaves rising to 24in/60cm. In late spring the fruiting stem appears among the leaves, a pale green pointed shoot which ripens in late summer to a sceptre, 6in/10cm long, of gleaming orange-red fruit packed together – curious and cheerful but not beautiful. Throughout the winter and spring the foliage provides a richly patterned companion for other

Illustration opposite:
Asphodeline lutea

plantings – snowdrops and *Cyclamen coum* in the late winter, followed by *Helleborus orientalis*, Solomon's Seal and lily-of-the-valley. In my garden it grows in dry shade, under the branches of an old yew tree, with ferns and Solomon's Seal, but it will also do well in part shade. It may be propagated by seed or by dividing the tubers after flowering.

Asphodeline

Asphodels in Greek mythology were connected with the Underworld and Homer describes a field of them that were the abode the dead. It is puzzling that such a cheerful plant could have such associations. There are about 20 species, all rhizomes, in the family Asphodelaceae/Liliaceae native to the Mediterranean area and to Asia Minor.

Asphodeline lutea
Origin: Mediterranean
Z: 7

❧ The yellow asphodel, or king's spear, is among the most exotic bulbous plants of the late spring. Its tall stems rise 36in/90cm and are crowned by a mass of flowers tightly packed into a bottle-brush shape. They are golden yellow with long whiskery yellow stamens. The leaves are very decorative – wiry and twisting, pale and glaucous, they writhe about the feet of the flowering stems. It requires a sunny position and, although it needs moisture at flowering time, it does not like waterlogged soil in the winter. It may be propagated by dividing the rhizomes in autumn – the silvery new rosettes of foliage are very attractive.

Asphodelus

The Greek *asphodelos* became in English 'affodil', and eventually 'daffodil' with which, botanically, it has little connection. There are 12 species of asphodel, all rhizomes, in the family Asphodelaceae/Liliaceae.

Asphodelus aestivus
Origin: N. Africa, Canary Islands, S. Europe
Z: 8

❧ This spectacular and lovely plant is, for some reason, rarely seen in gardens. Also known as *A. microcarpus*, it has a habit of growth unlike other asphodels. It has a branching growth, with flowering stems emerging alternately along the upper part of the

Illustration:
Asphodelus aestivus

central stem, which rises to a height of about 4ft/1.2m,
making the whole plant in flower resemble an exotic
plume. The flowers are arranged in upward-pointing
racemes, each flower bud white and striped in brown.
They open in June, white six-pointed stars with a stripe
down the centre of each petal and orange-yellow
anthers carried on very thin stamens so that they look
like tiny insects hovering at the mouth of the flower.
The leaves are rather lax, glaucous-grey, and form a
sprawling mound from which the flowering stem
erupts. It must have sharp drainage and a sunny
position. It is a wonderful border plant, throwing up a
delicate veil of flowers, but at the same time having
statuesque presence. It would make a marvellous plant
to dominate a sunny corner or for repeat planting to
give splendid harmony to a large border.

Asphodelus albus
Origin: Europe
Z: 6

❧ The European native asphodel is found in the wild in high places up to 6,000ft/1,800m, often in meadow land. Its tall upright racemes of flowers, up to 36in/90cm high, have a splendidly exotic air. In the garden the flowers will appear in May, the long racemes crowded with white or the palest pink star-shaped flowers with lemon-yellow anthers thrusting far out. They are followed by caramel coloured seed-pods, glistening like a freshly-sucked sweet. The foliage is grass-like, glaucous-grey and of rather lax habit. It needs a sunny position in light, well-drained soil. It may be propagated by dividing rhizomes in the autumn or by seed. The white asphodel has an airy lightness that makes it an admirable ingredient for a white border with such plants as the white variegated honesty *Lunaria annua* 'Alba Variegata', late-flowering tulips such as 'White Triumphator' and off-white *Dicentra* 'Langtrees'.

Bulbocodium

There are two species of bulbocodium, both corms, in the family Colchicaceae/Liliaceae.

Bulbocodium vernum
Origin: Europe, Russia
Z: 4

❧ Flowering in February or March this little corm provides a dazzling splash of colour. The flowers which erupt from the ground, virtually without stems, are, in the best clones, a lively rosy-purple. Other clones have rather wishy-washy mauve or white flowers. The flowers open wide in the sun with the pointed petals well separated. They are now more than 4in/10cm high and the leaves, appearing just after the flowers, rise a little higher. In the wild it is found in alpine meadows and in the garden should have an open, sunny site with good drainage. It is easily propagated by dividing the corms. It looks very beautiful among paler crocuses such as *Crocus tommasinianus*.

Illustration:
Camassia cusickii

Camassia

The word camassia comes from the North American Indian name, quamash, which is still a widely used common name. There are about five species, all bulbs, in the family Hyacinthaceae/Liliaceae native to America. All those described below may be propagated by dividing the bulbs in the autumn. The species may be raised from seed.

Camassia cusickii
Origin: N.W. U.S.A.
Z: 5

❧ The form of the flowers of this camassia distinguishes it from other species. The petals are very narrow, and the flowers are profusely borne in a thick upright raceme as long as 15in/35cm – like the most elegant bottle brush that you ever saw. The flowers open in May, the palest possible blue, set off by crisp lemon yellow anthers. The clouds of pale flowers give the plant a wonderful lightness. It looks beautiful among the glaucous foliage and burgeoning flower heads of *Euphorbia characias* ssp. *wulfenii*. *C. cusickii* 'Zwanenburg' is a cultivar with much larger, deep blue flowers which are rather ponderous, much less distinguished than the type.

Illustration: *Camassia leichtlinii* 'Electra'

Camassia leichtlinii
Origin: W. North America
Z: 3

❧ This is a stately plant whose flowering stems rise as high as 4ft/1.2m. It flowers in early May with long upright racemes of flowers which may be as long as 12in/30cm. The individual star-shaped flowers are violet or blue but the best clones are three particularly beautiful cultivars. *C. leichtlinii*. 'Alba' is one of the best white-flowered bulbous plants. Each flower is up to 2in/5cm across, of a lovely silken texture delicately lined with veins. The dark crescent-shaped anthers stand out prominently. *C. leichtlinii*. 'Electra' was an introduction of the nurseryman Eric Smith and has even larger flowers of an especially good rich lilac. In the border it has all the presence of some much

Illustration opposite:
Camassia leichtlinii
'Semiplena'

later-flowering herbaceous plant such as delphiniums.
C. leichtlinii 'Semiplena' is a very beautiful double
cultivar with beautiful creamy white flowers – a
marvellous ingredient for a border with a pale colour
scheme. *C. leichtlinii* needs rich feeding and will flower
well in a partly shaded site.

Camassia quamash (syn.
Camassia esculenta)
Origin: W. U.S.A.
Z: 5

❧ This is a variable plant but never less than
distinguished. From a mound of shining narrow
strap-like leaves the flowering stems emerge in April.
The stems are up to 36in/90cm tall surmounted by an
upright raceme. The lower buds start to open first,
revealing star-like flowers with long stamens with
sprightly yellow anthers. Different clones will have
flowers that range from deep rich blue, through a paler
sky-blue to pristine white. It is a superb border plant,
associating admirably with other spring flowering
bulbs such as tulips. In the wild it is widely distributed,
being found both in damp meadows and quite dry
places. In the garden it will do well in semi-shade and
would be an excellent late-flowering spring plant for
the meadow garden or orchard where it will rise above
even quite long grass.

Illustration: *Canna*
'Wyoming'

Canna

There are about 10 species of canna in the family Cannaceae. They are all rhizomatous perennials native to Central and South America. These splendid statuesque plants are usually thought of as tender bedding plants. Some, in fact, are hardy in many gardens in most winters but it only needs one really severe bout of deep frost to kill them. Also, they really demand favourable conditions to attain their full size, so the site you choose must be warm, protected and sunny. They are not in any case for timid gardeners, and are too wild and wonderful for the polite border. But in some exotic, jungle-like arrangement they provide in late summer an excitement in leaf and flower which little else can equal in that season. All have bold, upright foliage and are crowned by plumes of flowers which are small in relation to the whole plant but sometimes of brilliant colouring. The rhizomes should be lifted before the winter and stored in slightly moist compost in some frost-free place. They can then be planted out in late spring when the ground has warmed up. They need plenty of nourishment and water to shoot up to their full decorative size in the short growing season available to them in most gardens. Cannas may be propagated very easily by dividing the rhizomes in the spring.

Most cannas available are cultivars which have arisen from complicated hybridisation. However, some of the species and their close relations still provide some marvellous garden plants. *C. indica* 'Purpurea' may grow as high as 6ft/1.8m and has striking deep red flowers. *C. iridiflora* from Peru, supposedly hardier than many, has delicately formed flowers with narrow petals in pink or rich orange yellow and very handsome foliage. *C. malawiensis* 'Variegata' has leaves that are striated in pale green and gold. The flowering shoots have a pink bloom and the flowers are apricot-orange. Of the fifty or so cultivars available commercially 'Louis Cayeux' is an excellent pale pink and 'Wyoming' is a vibrant orange with bronze-tinged foliage.

Cardiocrinum

Cardiocrinum giganteum
Origin: Himalayas
Z: 7

There are three species of cardiocrinum, in the family Liliaceae/Liliaceae.

❧ There is nothing at all like this giant, aristocratic bulb which takes years to produce its flowers and needs very special conditions. However, it gives unique and spectacular pleasure, well worth waiting for. Its flowers in late June are immense white trumpets, each up to 8in/20cm in length, opening out sharply at the tips revealing a plummy interior. The flowers give off a deep, rich, exotic scent which few garden plants can equal. It needs a shady position in rich, moist soil that is, nonetheless, well drained for in waterlogged places the bulbs, over a long period of time, will rot. In appropriate conditions it will, after six or seven years from sowing, throw out its immense, stiff flowering stem, rising as high as 12ft/3.6m, crowned with as many as 20 downwards sloping flowers. It is monocarpic (i.e. flowering once only), producing huge quantities of seed but mature bulbs will split and may be divided, These new bulbs will produce a flowering stem in about four years. Much nourishment is need to support these fast growing, giant stems and a deep mulch of well-rotted manure in the spring will help. They also require moisture at growing time. The only place for it is in a woodland garden in the large-scale context of trees and shrubs. In a planned border it must have substantial companion planting.

Chionodoxa

The name of these bulbs, of which there are six species, means 'glory of the snow'. They are in the family Hyacinthaceae/Liliaceae and all come from the Mediterranean region.

Chionodoxa luciliae
Origin: Turkey
Z: 4

❧ The blue glory of the snow is one of those miniature plants, in itself without tremendous presence, which can enliven a corner or, planted in quantity, produce a dazzling splash of colour. It grows 6in/15cm high and the flowers in February or March are a sprightly clear blue fading to a chalk-white centre with a pale yellow eye. The flowers are star shaped, with pointed petals, held above fresh green pleated leaves. It will flower well in part shade, the sprightly blue showing to great advantage. In the wild it is found in stony, poor soil in harsh mountain conditions. In the more benign environment of the garden in a temperate climate it will establish itself vigorously. It may be propagated by dividing clumps of bulbs in the autumn or after flowering, or by sowing seed. It makes an excellent underplanting for deciduous shrubs such as magnolias. The first flowers of the early-flowering *Magnolia stellata* will overlap with the last flowers of *Chionodoxa luciliae*. It also associates well with other spring bulbs – with the paler yellow daffodils, for example. The slightly later-flowering *C. forbesii* is similar, with deeper blue flowers; there is a very unsatisfactory blotchy pink cultivar, *C. forbesii* 'Pink Giant', which sounds as horrible as it looks.

Colchicum

There are over 40 species of colchicum, all corms, in the family Colchicaceae/Liliaceae very widely distributed in Asia, Africa and Europe. They are often misleadingly referred to as 'autumn crocuses'.

Colchicum agrippinum
Origin: Garden
Z: 6

❧ This exquisite colchicum provides a dazzling sight in September. The flowers are exquisitely formed – rosy-purple trumpets of long, narrow, pointed petals held triumphantly at the tips of pale stems. The petals are speckled with vague chequered spots and the anthers are a rich purple. It rises no more than 4in/10cm and in the right conditions (not hard to provide) it will form generously floriferous clumps. The narrow, pointed leaves appear well after the flowers. It is unknown in the wild: its origin is uncertain but it is thought to be a hybrid of *C. autumnale* and *C. variegatum*. It is best in a well drained site and it will flower well in dappled shade. I have seen it looking beautiful with the newly opened foliage of *Cyclamen hederifolium*. It is easily propagated by dividing the corms in the spring or by sowing fresh seed.

Colchicum byzantinum
Origin: Garden
Z: 6

❧ This is one of the oldest garden flowers, being known to the great botanist and gardener Carolus Clusius at the beginning of the 17th century. It flowers in September, an admirable pink-purple, with much darker stamens. It is very vigorous and will establish itself easily bringing the atmosphere of spring exuberance to the autumn garden. It rises to about 8in/15cm high and its flowers are neatly formed. The leaves, appearing well after the flowers and taller than them, are very striking, ribbed and up to 6in/15cm wide. There is a white cultivar, *C. byzantinum* 'Album'.

Colchicum speciosum
Origin: Caucasus, Iran, Turkey
Z: 6

❧ The appearance of this large colchicum in autumn is always an exhilarating surprise. As the rest of the garden is starting to go to sleep, its leafless flowering stems shoot through the earth unfurling their purple flowers in October which rise to 8in/15cm high. Their leaves, too, are a great surprise for they appear in the following spring, so large, broad and gleaming that you find it almost impossible to believe that they are related to the same plant. The flowers are among the most cheerful autumn plants, often to be seen thrusting through the fallen leaves of deciduous shrubs and trees. The flower is 3in/8cm long and remains neatly cupped, with the petals opening out only slightly in the brightest sunshine. It is a tough plant, growing in the wild in mountain pastures as high as 10,000ft/3,000m. In the garden it would make a marvellous meadow plant were it not for the dilemma of when to cut the grass. It is not difficult as to soil and will naturalise obligingly. There is a most beautiful white cultivar, *C. speciosum* 'Album', which flourishes in my heavy clay.

Convallaria

This is one of those unusual genera of which there is only a single species, in the family Convallariaceae/Liliaceae.

Convallaria majalis
Origin: Northern Hemisphere
Z: 3

❧ The lily-of-the-valley has the most delicious scent of any garden plant, with a piercing freshness which is irresistible. For that reason alone it is an essential presence in the garden. It is also very easy to please – it even grows through the tarmac in my yard. The emerging shoots look very distinguished, fleshy and suffused with a grey bloom. The leaves are oval and crisply pointed, an attractive glaucous green, pleated

down the centre and marked with fine striations. The
flowering stems are 6in/10cm long and the flowers
which appear in May are creamy white, diminutive
bells the tips of whose petals reflex backwards. The
flowers are generously borne, up to a dozen on each
arching stem. In the wild this is a plant of woods and
meadows, found growing as high as 8,000ft/2,400m. In
the garden it will flourish in the shade, relishing a cool
position. I grow it in rather heavy soil, which seems to
please it, among ferns, hellebores and *Brunnera
macrophylla*. I have seen it grown in pots and brought
indoors to flower early and fill a room with its lovely
scent. Flat-dwellers with window boxes may grow it
successfully, with the perfume wafting in through the
open window. There are various cultivars of which
C. majalis 'Fortin's Giant' is simply a larger version of
the type – almost twice its height. The variegated form,
C. majalis variegata, always seems to me a rather feeble
little plant. Lilies-of-the-valley are very easily
propagated by dividing the rhizomes in the autumn.

Corydalis

The name corydalis comes from the Greek for the crested lark which is said to resemble the shape of the flowers. There are about 300 species, in the family Papaveraceae. They are all herbaceous perennials some of which are rhizomes or tubers.

Corydalis flexuosa
Origin: W. China
Z: 7

❧ Few plants have spread so quickly into gardens as this dazzling newcomer which was introduced into cultivation in the U.S.A. in 1987 and in Europe in 1989. It spreads by rhizomes, forming a vigorous mound of beautiful, fern-like foliage about 6in/15cm high. Each leaf is finely cut, pale green, with a reddish base and spots on the leaf. The flowers appear in April and continue for weeks, producing flowers intermittently throughout the season. They are held handsomely above the foliage, a wonderful rich blue with a hint of green. Four clones show slight but attractive variations: *C. flexuosa* 'Purple Leaf' has striking purple-bronze foliage and a rich violet-blue flower; *C. flexuosa* 'Père David' has paler blue flowers; and the foliage and flower buds of *C. flexuosa* 'China Blue' are flushed with pink. A fourth cultivar, *C. flexuosa* 'Blue Panda',

Illustration: *Corydalis flexuosa* 'Père David'

was collected by the American nurseryman Reuben Hatch who was responsible for first introducing the plant into cultivation in 1987. This has especially vivid blue flowers, with a characteristic upward-pointing curved spur, and bright green finely cut leaves. In the wild *Corydalis flexuosa* is found in humus-rich moist soil in woodland. I grow it close to a high north-facing wall in heavy soil, which seems to suit it. The flowers are produced over a long period – showing an especially intense colouring in cold weather. The foliage remains ornamental throughout the season. It makes a beautiful underplanting to hellebores which enjoy similar conditions. It is easily propagated by dividing the rhizomes in late summer or in autumn.

Corydalis lutea (syn. Pseudofumaria lutea)
Origin: Europe
Z: 6

❧ This European native is the type of easy plant – some say too easy – which inconspicuously adds character and visual texture to the garden. It has very ornamental foliage, an excellent glaucous grey-green, with finely cut and lobed leaves. The little flowers in May are a sharp lemon yellow, shown off strikingly by the foliage. In the wild it is found in rocky limestone country but I find it grows well in my rich, moist soil, ornamenting shady corners. It spreads by rhizomes by which it is easily propagated. There is a white form, *C. lutea* 'Alba', but far better is the white-flowered

Illustration:
Corydalis lutea

C. ochroleuca (also known as *Pseudofumaria alba*),
which is fibrous-rooted, with delicate putty-white
flowers with spots of green on the lip and lemon
throats. The foliage is particularly finely cut and the
fleshy shoots flushed with dusty pink. It grows well in
the shade and makes an admirable partner for all sorts
of other shade-loving perennials. Both will form
burgeoning mounds, up to 10in/25cm high. They will
seed themselves gently, showing a particular liking for
the limestone walls of my garden where they attach
themselves, hanging in splendid swags.

Corydalis solida
Origin: Asia, Europe
Z: 6

• With its upright racemes of purple flowers in April
this tuberous corydalis is a cheerful native plant of
hedgerows and woodland found very widely in
Europe. It has the decorative finely-cut leaves of most
of its tribe. The wild type would scarcely earn its place

Illustration: *Corydalis solida* 'George Baker'

in the garden, except in a wild corner. But a marvellous cultivar, *C. solida* 'George Baker', is entirely garden-worthy. It has flowers of a rich, warm salmon-pink which are beautifully displayed against the glaucous foliage. It forms a plant no more than 6in/15cm high and, is an excellent subject for the alpine garden or rockery. I have seen it looking distinguished overflowing the rim of a terracotta pot. It must have a well-drained position.

Cosmos

There are about 25 species of cosmos in the family Compositae. They include those useful half-hardy annuals with finely cut leaves which are cultivars of two Central American species *C. bipinnatus* and *C. sulphureus*.

Cosmos atrosanguineus
Origin: Mexico
Z: 8

❧ This tuber produces flowers of an extraordinary colour – a very deep chocolate-purple, almost black when they first open. They appear in June or July and carry on for many weeks, deep into autumn. They are 1 1/2in/4cm across and give off a sweet, vanilla scent with overtones of chocolate. It has attractive deeply cut foliage and the finely hairy stems are flushed with purple. It absolutely needs rich well-drained soil and plenty of sun. The tuber is relished by slugs, and many plants have been eaten in the heavy slug-laden soil of my garden. However, it is a magnificent plant for pots, making a marvellous contribution to an arrangement of red and purple. It is also very easy to propagate, by semi-ripe cuttings in later summer, so in less benign gardens it may be treated as a tender bedding plant.

Illustration:
Crinum × powellii

Crinum

Crinum bulbispermum
Origin: South Africa
Z: 6

There are well over 100 species of crinum, in the family Amaryllidaceae/Liliaceae, all bulbs and distributed widely in the tropics and the southern hemisphere. Only a handful are available commercially and these tend to be species that are hardy enough to survive in the garden in warmer places.

⁊ This species crinum is, with *C. moorei*, one of the parents of the hybrid *C. × powellii* described below. It has a strong character all of its own and is well worth seeking out. The flowering stems rise to 4ft/1.2m and the flowers open in late July or August. These are pale pink trumpets, smudged on the backs of the petals with rosy purple, and giving off a delicious sweet scent. The flowers are most elegantly held, curving slightly downwards on long arched stems. The strap-like leaves are a decorative glaucous green. It must have a sunny position and rich soil. Like *C.* × *powellii* it seems to flower best when the bulbs are fairly constricted. It makes a magnificent bulb for a substantial pot.

Crinum × *powellii*
Origin: Garden
Z: 6

&❧ Few bulbous plants have such presence in the garden. Masses of leaves are produced, broad and glistening, from which the thick flower stems, rising 36in/90cm, emerge to produce their flowers in August. These are wonderfully elegant, with the buds curving out from upright flower heads, resembling the beak of an exotic bird before they open. The flowers are the palest pink, gently trumpet-shaped, with the petals curving crisply backwards at the tip. The petals have the texture of silk and the stamens with their curving anthers crowd the throat. The flowers give off a sweet scent and last for many weeks making one of the most exciting plants that performs late in the season – I have seen them still in flower in October. To flower well it needs a sunny position. The books say that it needs sharp drainage but I have grown it successfully in heavy clay. The bulbs should not be planted too deeply – their necks should just break the surface. It is said to flower best when the bulbs are packed together or restricted within the confines of a pot. The white cultivar *C.* × *powellii* 'Album' is, if anything, even more beautiful than the type.

Crocosmia

There are about six species of crocosmia in the family Iridaceae all native to South Africa. They were formerly known as montbretia. They interbreed with great ease so that the true identity of both species and cultivars is difficult to establish. However, they are essential garden plants, easy when established and ornamental both in foliage and in flower. The flower

Illustration: *Crocosmia*
'Golden Fleece'

juts sideways, a spike of flowerlets whose petals curve
backwards, in many cases thrusting the stamens far out.
They will flower well in part shade or full sun and are
at their best in moisture-retentive soil, needing water as
the flowers develop. In British gardens it is striking
how well they do in the wet west of England or on the
west coasts of Ireland and Scotland. Most are hardy to
Zone 7 but in many colder places if the corms are
planted deep and perhaps mulched in the winter they
will have a good chance of surviving. The species, such
as *C. masoniorum* and hybrids like *C. × crocosmiiflora*
are pretty enough but the most valuable garden plants
are certainly to be found among the various cultivars.
C. 'Lucifer' is a magnificent July-flowering variety with
vibrant scarlet flowers and particularly handsome
foliage, dark green, quite broad and upright. *C.*
'Rowallane' is similar in stature, also flowering in July,
with very large pale apricot flowers. *C.* 'Emily
Mackenzie' is much later flowering, in August and
September, with excellent rusty-orange flowers
splashed with darker spots within the flowers. It grows

Illustration:
Crocosmia 'Lucifer'

to about 24in/60cm tall, delicately formed. C. 'Golden Fleece' (formerly C. 'Citronella') is another of the smaller kinds but with beautiful pale yellow flowers, appearing late in the summer and often continuing into the autumn – as late as October in my garden. C. 'Solfatare' is another yellow-flowered variety that flowers in late summer and autumn. C. 'Red Knight' will grow as tall as 36in/90cm, with marvellous deep red flowers in August or September.

Crocosmias make a contribution to the late summer border which few other plants equal. Even before their flowering period the stiff, blade-like leaves are an attractively firm shape in a densely planted border. In flower, with their shades of yellow, apricot or red, they give fresh colour when other herbaceous plants are flagging. The reds and oranges are superlative as ingredients in a hot-toned border. They are magnificent with alstroemerias, dahlias, daylilies and appropriate bedding plants, perhaps set against a background of the deepest purple foliage of *Cotinus coggygria*. The yellow-flowered kinds look good with orange-flowered plants such as *Lilium pardalinum* or *Alstroemeria aurea* or with the blues of delphiniums. Crocosmias are easy to propagate by dividing the corms which can become very congested.

Crocus

There are about 80 species of crocus, all corms, in the family Iridaceae native to mid Europe, the Mediterranean region and Central Asia – in a very wide range of habitats from the banks of the Mediterranean to the mountainous regions of Afghanistan. For the garden the most precious species are those that will establish trouble-free colonies. Some, too delicate and too demanding in terms of site for life in the border, are only appropriate for the Alpine house or trough.

Crocus gargaricus
Origin: Turkey
Z: 7

This is not the toughest of crocuses but it is so beautiful that it is worth any trouble taken in its cultivation. It flowers in March, ghostly shoots, the palest creamy white, thrusting from the soil from which the rounded flower heads emerge. These are the loveliest warm golden yellow, finely veined in a darker shade. In full sun they open out, showing their pointed petals. In the wild *C. gargaricus* grows in meadows, up to 10,000ft/3,000m. In the garden it should be given a well-drained site, in the sunniest place you can find. In a cold-frame or unheated greenhouse it is easy to provide the perfect conditions, bringing the pot into the house when flowering is about to start, to provide one of the most beautiful house-plants imaginable.

Crocus goulimyi
Origin: Greece
Z: 7

❧ This is one of the most beautiful of the autumn-flowering crocuses. Its flowers, which appear in September or October, are sweetly scented and vary from pale to richer lilac, borne on pale stems which rise 6in/10cm high. The petals are marked with finely etched lines in deeper purple and the anthers are egg-yolk yellow. In the wild it is found in rocky places in limestone country. In the garden it should be given a sunny site and appears to flourish in all but the most waterlogged soil. It may be propagated both by seed and by division of corms. There is a beautiful white cultivar *C. goulimyi* 'Mani White' (syn. *C. goulimyi* 'Albus').

Crocus minimus
Origin: Mediterranean
Z: 8

❧ This enchanting little crocus is not a plant to make much of an impact in the garden. But it is very beautiful and it is well worth trying to find a position that will provide the conditions it needs. Its flowers, which appear in February or March, are a rich lilac colour, opening quite flat in the sun, and with rounded petals. The petals are marked with a deeper purple feathering, most striking on the back, and the stamens are deep golden yellow. The flower stems rise no more than 2in/5cm and the narrow grass-like leaves are much taller, half concealing the flowers. In the wild it is found in sandy soil sometimes, in Corsica or Sardinia, at the edge of a beach. It is a plant for the rock garden or for a pot or trough raised up on a sunny terrace where its exquisite flowers may be admired close-up.

Crocus sieberi
Origin: Greece
Z: 7

꙰ This is not the hardiest crocus but it is one of the prettiest. The flowers in February or March are pale lilac – there is a distinguished deeper purple form *C. sieberi* var. *atticus.* All have a striking orange-yellow throat to the flower and narrow grass-like foliage. In its native habitat *C. sieberi* is found in the mountainous regions of Greece, exceedingly dry in the summer. In cooler countries it must have a sunny position and good drainage. I grow it at the foot of the common rosemary (*Rosmarinus officinalis*) in a south-facing border. Other small shrubs of a Mediterranean character – artemisias, lavender and sages – will provide good companion plants which also enjoy the same conditions. *C. sieberi* 'Bowles White' is a beautiful pure white cultivar.

Crocus tommasinianus
Origin: E. Europe
Z: 5

꙰ In early February the pale mauve flowers of *Crocus tommasinianus* are an exhilarating sight. Up to 4in/10cm high, they are the palest grey-mauve on the outside, a deeper violet within, with striking yellow

stamens which give off a powerful smell of saffron in the sun. The leaves are very narrow, almost like grass, marked with a pale stripe that emphasises their elegance. All gardeners should try and provide a site for this marvellous crocus that will allow it to multiply and thrive. In its natural habitat it is a woodland plant and in the garden should have at least semi-shade. I have it growing under a small ornamental deciduous tree, *Amelanchier canadensis,* which seems to give it the conditions it needs, providing summer dormancy in dry soil in the shade of the tree. It seeds itself lavishly, even spreading into the adjoining lawn. I have seen it flourishing in informal areas of grass which should have been cut as late as possible to allow the crocus to be visible. It associates well with other spring-flowering plants such as snowdrops, aconites and *Anemone blanda.* There are some good forms, including a lovely pure white one, *C. tommasinianus albus.* There are several cultivars which add nothing to the quality of the type.

Crocus vernus
Origin: E. and S. Europe
Z: 4

ᵛ✿ The species has a slightly cupped, rounded flower, mauve or white, up to 4in/10cm high. It flowers in February or March and is very rarely seen in its wild type, for it is the origin of countless cultivars which provide plants for large-scale planting in public gardens. Individually many of these cultivars are rather coarse but *en masse,* especially with carefully chosen combinations of colour, they can be very effective. In short grass, under deciduous trees, they will make a dazzling spring spectacle.

Cyclamen

There are about 20 species of cyclamen, all tubers, in the family Primulaceae. The word comes from the Greek *kyklos,* a spiral, referring to the twisting stem of the seed pod. They come from Europe and the Mediterranean area, with one species from central Africa. The tubers may be bought when dormant or when in leaf. Different species produce their roots from different surfaces of the tuber so, if planting when dormant, it is best to plant it on its side – it will settle into the correct position of its own accord. Always buy a generous quantity and plant them in a substantial group – one or two merely look absurd. In appropriate conditions all the cyclamen described below will self seed easily. In some years this will be so prolific that it is best to transplant seedlings to less crowded areas. Despite their delicate appearance they are very tough plants and among the most versatile. Different species will flower in winter, spring and late summer and in each case their foliage continues to provide ornament long after the flowers have gone.

Cyclamen coum
Origin: Caucasus
Z: 6

ᵛ✿ The winter-flowering cyclamen provides a splash of brilliant colour given by no other hardy plant in that season. The flowers in January are a cheerful purple-pink, raised well above the foliage. Before they open, the buds curve downwards elegantly from their red-brown stems. In a community of plants the white form frequently appears creating a striking effect like some summer fruit splashed with cream. The leaves are

Illustration opposite:
Cyclamen coum

rounded and of a leathery texture with handsome markings: at the centre of each leaf there is a darker shape of an ivy-leaf edged with a vague silver outline. In young plants the underside of the leaf – which you will never see unless you turn it over – is a rich maroon. It will flourish in shade or part-shade but I have seen it well established in the short grass of a sunny orchard. It flowers in the same season as other small bulbous plants such as snowdrops and winter aconites (*Eranthis hyemalis*) which together make a brilliant winter picture. There are several selected forms, those with especially pale, silver colouring to the foliage being especially appreciated. *C. coum* 'Maurice Dryden' is a beautiful one with white flowers. There is a particularly attractive form with a clear pink flower, *C. coum roseum.*

Cyclamen hederifolium (syn. C. neapolitanum) Origin: S. Europe, Turkey Z: 6

❧ The so called ivy-leafed cyclamen produces leaves that roughly resemble those of ivy but they are very variable and, in their marking, far more intricate and beautiful than any ivy. The flowers appear in late summer just at the moment when most other perennial plants seems to have produced their last flowers. The flowers are diminutive – no more than 3/4in/2cm long – like little sails, held aloft on brown-tinged stems.

They vary in colour from a cheerful purple pink – sometimes attractively astringent – to pure white. In late autumn the spherical seed-heads are held at the tips of intricately coiled stems. Before the flowers have faded the leaves start to unfurl revealing the most decorative aspect of this cyclamen. The leaves are bold, up to 4in/10cm long, and varied in shape – some pointed, some almost circular and others heart-shaped or finely scalloped. They have deep veins and are dappled with marbling. Their colour varies from deep green to a much lighter colour of pale silver green. In a vigorous clump the leaves will crowd together and overlap making a striking pattern. They remain throughout the winter, forming a wonderful background to winter-flowering plants such as snowdrops, cyclamen and aconites. They continue into the summer, falling only in June. In the wild it is found in shady places, in woodland or in olive-groves. In the garden it will do well in a north-facing position or scattered round the trunk of a deciduous tree or shrub.

Cyclamen purpurascens
Origin: Europe
Z: 6

❧ This cyclamen produces its flowers in late summer and early autumn. They are a lively carmine pink with a much darker patch at the base of the flower. The flowers are deliciously scented and a substantial colony will suffuse the air with its perfume. The leaves are rounded or gently heart-shaped, up to 3in/8cm across, with purple markings on the underside and decorative marbling on top, often with a pale margin surrounding an ivy-leaf shape at the centre. In the wild it is found in rich calcareous soil and in the garden is best in moist soil in dappled shade. It is beautiful among ferns and hellebores at the edge of a shady woodland garden.

Cyclamen repandum
Origin: Mediterranean
Z: 7

❧ This is the spring-flowering counterpart of *C. hederifolium* with distinctive virtues of its own. Its flowers appear in April, a sprightly pink-magenta, with twisted petals resembling a rather muddled ship's propeller, and marked with a splash of purple at the base. The flowers are deliciously scented – the perfume very striking on warm spring day. The flowers are held

well above the foliage on red-brown stems and the
flowering shoots will scramble through low-growing
plants scattering brilliant dabs of colour. The leaves,
appearing at the same time as the flowers, are scalloped,
finely toothed, loosely heart-shaped and delicately
marbled with paler green. At the centre of each leaf
there is the darker shape of an ivy-leaf. They are just as
beautiful as those of *C. hederifolium* but do not have
the same decorative impact, being soon overwhelmed
by the rush of summer plants. In the wild it is a
woodland plant, often found in quite dry places in pine
forests. In the garden it is very versatile but always at
its best with some shade. The flowers appear at a time
when yellow, blue or violet are the dominating colours.
It looks beautiful among ferns or with snake's head
fritillaries (*Fritillaria meleagris*), erythroniums, wood
anemones (*Anemone nemorosa*) and the pale violet
Scilla messeniaca all of which flower at the same time. I
have also seen it intermingled with inky-purple grape
hyacinths (*Muscari armeniacum*), like a spread of rare
mixed jewels. It may be grown successfully in a
meadow garden. A subspecies, *C. repandum* ssp.
peloponnesiacum, has richer carmine flowers and paler
markings on the leathery foliage – creating a most
beautiful effect.

Illustration: *Dactylorhiza praetermissa*

Dactylorhiza

There are about 30 species of dactylorhiza, all tubers, in the family Orchidaceae. All genera of Orchidaceae are protected by the Convention on International Trade in Endangered Species (CITES). Responsible nurseries will propagate these plants from seed gathered in the wild. Responsible gardeners will make sure they obtain their plants *only* from such nurseries.

Dactylorhiza elata
Origin: Mediterranean
Z: 6

This great marsh orchid produces spectacular flowers carried on stems that may rise as tall as 36in/90cm. The flowers appear in May or June borne on a cylindrical head up to 24in/60cm long. Each flower is exquisitely shaped, like an exotic bird in flight, deep violet purple with a darker centre and spots in the throat. In the wild it is found in damp places and in the garden it should have such a site, perhaps on the banks of a stream or pool, in an open situation or one of dappled shade. Such wild plants are best in a naturalistic setting and strikes a jarring note in an artfully contrived border.

Dactylorhiza praetermissa
Origin: N.W. Europe
Z: 6

❧ The southern marsh orchid is found in the wild in marshlands on calcareous soil. It has an imposing flowering stem rising as high as 18in/45cm crowned in June with a dazzling spike of tightly packed flowers. These are purple-red with intricate winged petals splashed with spots of deeper colour. The glaucous leaves form a handsome plumed shape at the base of the flower stem. It may be propagated by seed or by division in the early spring. This is a plant for the wilder parts of the garden. Plant it in dappled shade on the grassy bank of a stream or pool to follow primroses and cowslips.

Dahlia

There are about 30 species of dahlia, mostly tubers, in the family Compositae, all native to Central America. Apart from the one or two species described below, most garden dahlias are the highly selected hybrids of *D. coccinea* and *D. pinnata*. There are well over 400 different cultivars available commercially, and very many more have been lost to cultivation. The International Register of Dahlia Names carries the names of 20,000 cultivars. From their introduction early in the 19th century they quickly became popular. Cannell & Sons' nursery in Kent offered for sale 469 cultivars in 1890. They have remained collector's plants, with societies specialising in them, but polite garden taste has spurned them in recent times. However, with an interest in bolder schemes, their qualities are once more being recognized. Many of them provide a richness of colour and an irresistible flamboyance of flower that makes them brilliantly ornamental plants in the right setting. The National Dahlia Society of England has devised a classification based on flower shape which divides them into 31 categories, minutely defined.

The cultivars described below will start to flower in July or August and continue for many weeks, often continuing until the first winter frosts. They are tender, reliably hardy only in Zone 9. In less favoured gardens the tubers may be dug up when the first frosts have

Illustration oposite:
Dahlia 'Arabian Night'

Illustration: *Dahlia* 'Bishop of Llandaff'

blackened the top growth and then allowed to dry off in a frost-free place. They may then be stored, again in a frost-proof place, in clean soil which will prevent them from drying out entirely. When growth starts in the spring they should be transferred to richer compost before planting out when there is no danger of frost. A less laborious alternative in colder gardens, and those with heavy soil, is to keep the tubers in pots and bed them out, pot and all, in the spring, to be removed in late autumn and kept in a frost-free place.

They range very widely in colour and flower type, some are single-flowered, many are elaborately double with hundreds of petals forming a huge pompom. Some have flowers only 2in/5cm in diameter, others, especially those grown for competitions, have giant, sometimes rather absurd, flowers as much as 12in/30cm across.

D. 'Angora' is very double, white, with petals that are crimped at the tips, giving the impression of a powder puff. *D.* 'Arabian Night' has double flowers that are almost black when first open but fade slightly to a superb rich blood red. *D.* 'Bednall Beauty' has purple-bronze foliage and rich double red flowers. *D.* 'Bishop of Llandaff' (known in America as 'Japanese Bishop') is in the style of older kinds but, in fact, dates from 1928. It is a magnificent plant with strikingly

Illustration: *Dahlia*
'Chiltern Amber'

Dahlia coccinea
Origin: Guatemala
Z: 9

sombre purple-bronze foliage and large single flowers of an intense scarlet with bold yellow anthers and the texture of velvet. *D.* 'Easter Sunday' is white with single flowers with bold yellow centres surrounded by a secondary ruff of smaller petals. *D.* 'Yellow Hammer' is a good pale yellow with single flowers and rounded petals, resembling its parent *D. coccinea*. *D.* 'Chiltern Amber' is a good rich apricot colour. But cultivars come and go with bewildering speed and many, even when commercially available, are to be found at only a single nursery. It is probably best to get the catalogue of a specialist supplier and choose the colours you need for a particular scheme.

❧ This is one of the species of dahlias that have given rise to so many of the garden cultivars. It is in itself a magnificent garden plant, not, alas, very hardy, and can make a tremendous contribution to the border. In its native habitat it will grow as tall as 10ft/3m but it needs a long hot summer and plenty of feeding and water to do as well as that. Its flowers are large, single, a

dazzling pure scarlet with a crisp golden eye. The leaves are pinnate and the flowering stems are attractively flushed with purple. The plant has the distinctive dahlia characteristic of having hanging flower buds, looking curiously like greenish strawberries, which flop upwards when the flower opens, to present itself full face. It is magnificent among crocosmias, daylilies and other hot coloured plants. It must have sunshine and it will certainly need support.

Dahlia merckii
Origin: Mexico
Z: 8

❧ This species dahlia has none of the bravado of the large cultivars but it does have great charm and an admirably long flowering season. It will grow to 4ft/1.2m, throwing out airy branches with scattered flowers appearing from June onwards. These are single, lilac-coloured, with a yellow eye, and continue well into the autumn. It must have a sunny position in a very well-drained position. It is excellent in a mixed border where it will interweave with other plants. I have seen it used to great effect on a precipitous south-facing slope. There is a good white form, *D. merckii alba,* and an even better cultivar of it with finely pointed petals, *D. merckii* 'Hadspen Star.'

Dichelostemma

There are about six species of dichelostemma in the family Alliaceae/Liliaceae. They are all corms and native to the Western U.S.A.

Dichelostemma congestum
Origin: W. U.S.A.
Z: 8

❧ This charming plant shows its relationship to the onion tribe. It produces its flowers in late June, a rounded umbel of warm violet flowerlets with pale lemon centres. The umbels are held on rather stiff stems which rise as high as 24in/60cm. It should have a sunny position in very well-drained soil. It may seed itself, forming a colony, and looking very decorative among the smaller shrubs which enjoy the same conditions such as artemisias, cistus, lavender, sage and santolina. Its flowers are beautiful among the dusky foliage of *Salvia officinalis* 'Purpurascens'. It may also be propagated by dividing the corms in the autumn.

Dichelostemma ida-maia
Origin: W. U.S.A.
Z: 7

❧ The expressive common name for this cheerful plant is the firecracker flower. Its former scientific name was *Brodiaea ida-maia*. From a mass of leaves the wiry flowering stems, up to 12in/30cm tall, produce their dazzling flowers in late June. These are little crimson hanging tubes tipped with gold-green. They have a glistening texture, as though freshly painted. The flowers sway on their stems and provide brilliant spots of colour, looking especially beautiful among pale grey foliage. It must have a sunny position in well-drained soil – it will flourish in quite poor soil. It needs water in spring to develop its flowers but the corms welcome a completely dry position after flowering. It may be propagated by seed – in favourable conditions it will seed itself – or by dividing the corms in the autumn. It is an excellent plant for pots, alone or mixed with others, where it may be given precisely the conditions it demands. The corms may be lifted in the autumn and replanted in early spring in fresh compost with plenty of grit for good drainage.

Dierama

Dierama dracomontanum
Origin: South Africa
Z: 9

Dierama pulcherrimum
Origin: South Africa
Z: 7

There are over 40 species of dierama, all corms, in the family Iridaceae, native to central and southern Africa.

❧ This dierama is one of the less hardy kinds but it is a plant of tremendous character. It forms a clump up to 36in/90cm high and produces its flowers in June on stiff stems (unlike its swaying sister described below). The flowers are loosely trumpet shaped and range in colour from pinkish red to much deeper shades. The best colour is a deep, dusty red which would be a superlative ingredient in a border of hot colours. It is best in rich soil and needs plenty of moisture at flowering time. It has a long flowering season and looks beautiful in a border of greys, reds and purple with such plants as artichokes, geraniums, penstemons and sidalceas.

❧ The common name of this dierama, the wand flower, comes from the long, thin, swaying flower stems which rise like wands above the foliage. The flowers hang downwards on hair-like stems on which they swing gracefully. They open in June and vary in colour from pale rosy-pink to a much richer, almost magenta pink. The flowering stems will rise as high as 6ft/1.8m and the flowers are borne over several weeks. In the wild it grows in rich, moist soil and in gardens it

Illustration:
Dierama pendulum

is often planted in a sunny position on the banks of a stream or pool. In a border it is superbly ornamental, with the flowers suspended high above other plants. In rich soil it will form a substantial clump sending out generous flowering stems. The evergreen foliage is grass-like and rather scrappy and it is certainly no loss if it is concealed by other plants. It may be propagated by dividing clumps in autumn. Some particularly attractive cultivars have been bred by the Slieve Donard nursery such as *D. pulcherrimum* 'Blackbird' which has rich purple flowers. *D. pendulum* is very similar to *D. pulcherrimum* in all respects except that its petals stand out from the stems making the whole raceme fuller, forming a striking tassel.

Disporum

Disporum flavens
Origin: Korea
Z: 5

There are about 10 species of disporum, all rhizomes, in the family Convallariaceae/Liliaceae

❧ This decorative woodland plant has something of the character of a yellow-flowered Solomon's Seal (*Polygonatum*) but with distinct charm of its own. It produces its flowers in May, pale yellow hanging tubes with curiously prominent stamens which curl outwards at the tip. The flowers dangle attractively in a tuft of downward pointing leaves. The fleshy stems that carry both leaves and flowers rise about 12in/30cm high are

dark at the bottom becoming paler further up. The leaves are a fresh green, sweepingly oval with pointed tips and pronounced furrows on their surface. The leaves continue to be decorative long after the flowering is over. This is a plant for moist shade where its flower colour will associate particularly attractively with ferns that have yellow-green new foliage. It may be propagated by dividing the rhizomes in the autumn.

Dracunculus

Dracunculus vulgaris
Origin: Mediterranean
Z: 9

There are three species of dracunculus, all tuberous perennials, in the family Araceae, native to Europe.

❧ I can never quite make up my mind whether this is merely curious or actually decorative. It is certainly striking, and has handsome details, but is it garden-worthy? It is known as the dragon plant and has long been grown in gardens. Its new shoots are attractively mottled and its foliage boldly divided, rich green with pale markings. Its flowers in summer are a splendid sight – like those of a rakish cousin of 'Lords and Ladies' (*Arum maculatum*) – deep purple spathes with undulating margins with a long, pointed, almost black spadix. The flower gives off a horrible smell, like decaying flesh. The whole plant grows to a height of about 36in/90cm. It must have a warm, sunny position and may be propagated by dividing the tubers.

Epipactis

There are over 20 species of epipactis, in the family Orchidaceae, very widely distributed in Africa, America, Europe and Asia.

Epipactis palustris
Origin: Europe
Z: 6

❧ The marsh helleborine is a late-flowering orchid which would make a lovely ornament in the wild garden. It rises to a height of about 15in/35cm with, in late summer, a spray of flowers at the tip. These are very intricate, with wings resembling insects and coloured in shades of white, yellow and pink. The stems are rather hairy with upwards pointing leaves carried alternately. In the wild it is found in damp places in acid soil. But it will tolerate slightly alkaline soil provided it has sufficient moisture. It is a marvellous plant for the edge of a stream. It may be propagated in the spring by dividing the rhizomes.

Eranthis

The name *Eranthis* comes from the Greek for spring flower. There are about seven species, in the family Ranunculaceae. They are all tuberous-rooted perennials, native to Europe, Central Asia and Japan.

Eranthis hyemalis
Origin: S. Europe
Z: 5

❧ Few plants create a livelier scene than a fine spread of the winter aconite flaunting its gleaming, cheerful yellow on a January day – in a mild winter it may even be in flower before Christmas. The single flower is at first globular but opens out into a cupped shape, 1in/2.5cm in diameter, very much resembling its cousin the meadow buttercup (*Ranunculus acris*). The flowers

rise above a decorative ruff of leaves encircling the fleshy stems which are up to 2in/5cm long holding the flowers well aloft. After the flowers have fallen the leaves continue to grow, becoming larger and rising higher, making a decorative pattern. In the wild the winter aconite flourishes in humus-rich soil, neutral or calcareous, in woodland and scrub. In the garden it will do best in a similar position. An ideal position for it is at the foot of a deciduous tree where it will form a spectacular golden carpet, getting all the moisture it needs in the winter and lying dormant in the dry summer. In the woodland garden, or in some shady bed, it is an essential partner of other winter-flowering plants – snowdrops, Lenten hellebores (*Helleborus orientalis*) and *Cyclamen coum*. The foliage of other smaller plants such as *Arum italicum* ssp. *italicum* 'Marmoratum', pulmonarias and hardy geraniums make an ornamental background. A later-flowering cultivar, *E.* 'Guinea Gold' has deeper yellow flowers, set off by a ruff of leaves flushed with bronze, and a longer-flowering season. Propagation is by division soon after flowering, but it may take time for plants to settle down and flower successfully. This is a much more reliable way of acquiring than by buying the dried rhizomes which are sometimes offered for sale. Some failures are reported from gardeners who do not plant the rhizomes sufficiently deeply. If you acquire them 'in the green', which is best, make sure you plant them to exactly the same depth as is indicated by the plants, which will be between 2in/5cm and 3in/8cm of earth above the top of the rhizome.

Eremurus

There are about 40 species of eremurus in the family Asphodelaceae/Liliaceae.

Eremurus robustus
Origin: Afghanistan, Russia
Z: 7

❧ Few border plants combine to such a striking degree the presence and the delicacy of this lovely plant. Known as the fox-tail lily, its flower stems rise very tall, as high as 8ft/2.5m, bearing in June a long bottle-brush of flowers, sometimes long as 4ft/1.2m. The flowerlets are the palest pink, delicately formed with prominent stamens which give the whole surface a blurred texture. The foliage is strap-like and insignificant, often withering by the time the plant is in flower. It should have a sunny site and needs fertile, light soil. In windy places the flowering stems will be deformed, twisting hither and thither. At the back of a big border, rising airily above all other herbaceous plants, it is dazzlingly lovely. It has a long flowering season. It may be propagated by division.

Eremurus stenophyllus
ssp. *stenophyllus*
Origin: Central Asia
Z: 5

❧ This is a smaller fox-tail lily than *E. robustus*. Its flowering stems rise only to 5ft/1.5m making it less prone to wind damage. It is similar in other ways but the flowers are a pale yellow becoming a warm tawny gold as they age. It is a marvellous plant for the smaller border, especially with blues and golds. It is especially beautiful against deep purple foliage such as that of *Cotinus coggygria* 'Royal Purple' or of *Corylus maxima* 'Purpurea.' It may be planted in a very large pot but it should be accompanied by other planting at its base to conceal its rather scrappy foliage.

Erythronium

There are about 20 species of erythronium, all bulbs, in the family Liliaceae/Liliaceae native to North America, Europe and, a single species only, Japan. They are found in a very wide range of habitats from moist woodland to dry mountainsides. The name comes from the Greek for red which is the colour of the flowers of the native European *Erythronium dens-canis.* Some of the west-coast North American species are known in the U.S.A. as trout lilies and some, indeed, have leaf-markings that resemble the scales of a fish sparkling underwater. Although they vary in habitat they may all be propagated by dividing clumps of bulbs as the leaves die and replanting in enriched soil. In most gardens the best months will be June or July; in dry weather make sure that the new site is well watered. The colours of flowers and the markings of the foliage will vary within a particular species.

Erythronium americanum
Origin: Eastern North America
Z: 3

❧ This little erythronium has flowers that resemble miniature elongated daffodils of a lovely pale lemon yellow. The leaves are long and pleated down the middle, marked with handsome marbling of brown and grey. The flowers, which appear in April, are held on tall stems 4in/10cm high which curl over at the top. In its native America it is a plant of woodland and pastures and in the garden it looks beautiful in dappled woodland shade growing in rich moist humus among, for example, moss and wood anemones. Although the bulbs are quite small they should be planted fairly deeply – no less than 5in/12.5cm. Despite the delicacy of its appearance it is among the more robust of the erythroniums. Its great charm is the combination of sharp yellow with the simple form of its flowers rising above distinguished foliage. There are flashier plants but few more beautiful.

Erythronium californicum
Origin: California
Z: 5

🐦 This Californian erythronium grows in the wild in coniferous woods. It has striking flowers, opening in April from creamy pink buds. When the petals curve back they reveal a cream-yellow interior with protruding pale anthers. The interior is worth looking at – the base of the throat is marked with deep purple 'sight lines' to guide bees about their fertilizing business. Up to three flowers are borne on the tip of each stem which is dusky purple in colour and rises to a height of 12in/30cm. The leaves are especially attractive, a pale glistening green handsomely marked with paler marbling. They are rather more erect than in many erythroniums and form a decorative sheaf from which the flowering stems emerge. In the garden it will grow well in shade or part shade, performing best in rich but well-drained soil. *E. citrinum*, from California and Oregon, is very similar but with a sharper yellow to its colouring. 'White Beauty' is either a cultivar of *E. californicum* or a hybrid. It is more vigorous and very beautiful, forming generous clumps with the same decorative markings. The flowers have less yellow and are larger than the type. I grow it successfully in the shade, against a north-facing wall among the foliage of *Cyclamen hederifolium*.

Erythronium dens-canis
Origin: Asia, Europe
Z: 3

❧ The dog's-tooth violet – which is not a violet at all – takes its name from the curious shape of the bulb which slightly resembles a large, elongated tooth. It is the only native European erythronium and an exceptionally decorative plant. Its foliage is very striking, each leaf up to 3in/8cm long, rounded and pointed, and marked with deep purple marbling on a lively green background. The flowers, which open in March – the earliest of the erythroniums – are a sprightly purple-pink, trumpet shaped and at first downward pointing on curved stems. When they are fully open the petals sweep backwards revealing long pale stamens tipped with dramatic purple anthers. In the sun the colour of the flower will fade to a rosy pink. In the wild the dog's-tooth violet grows in woodland and meadows. In the garden it will flourish in the same sort of conditions that are best for cyclamen and hellebores – moist soil in part shade. After the flowers have faded the leaves will grow with greater vigour, making bold clumps and overlapping to form a lively pattern. The foliage looks splendid intermingled with that of *Cyclamen hederifolium*. There are several named cultivars few of which improve on the type and some of which are positively coarse and overblown – 'Pink Perfection', for example. A white variety, *E. dens-canis* 'White Splendour', has much charm, retaining the dark violet centre of the type.

Erythronium helenae
Origin: California
Z: 5

❧ The combination of highly decorative foliage and sparkling flowers makes this one of the finest erythroniums. Its flowers open in April, creamy white but with a rich lemon-yellow centre. They are held on dark brown stems 5in/8cm high which rise above exceptionally beautiful leaves. They are long, undulating and relatively narrow, marked with smudges of maroon-chocolate, and often crisply edged with the same colour. In the garden it should have a

partly-shaded site in well-drained but humus-rich soil –
it will not flourish in very heavy soil that retains
moisture in the winter. I have seen it looking
marvellous in the dappled shade of *Magnolia stellata*
whose flowers will appear at the same time – the pure
white contrasting with the cream and yellow of the
erythronium.

Erythronium hendersonii
Origin: N.W. North
America
Z: 5

❧ There is no such thing as an ugly erythronium –
this is an especially beautiful one. It flowers in April,
with the flowers on slender stems carried well above
the handsomely mottled foliage. The downward-
curving flower buds are a rich rosy-purple but when
they open they are much paler, with the petals fading
almost to white at the base. The petals separate and
curve backwards throwing the purple anthers into
prominence. It will bear as many as ten flowers on each
stem, up to 10in/25cm high, and the contrast of colour
between bud and flower makes a lively sight. In its
native California and Oregon it grows in clearings in
pine woods. In the garden it needs part shade and very
sharp drainage in a site that allows it to be almost
completely dry when dormant. The front of a sunny
border, among the sprawling branches of low-growing
shrubs of Mediterranean character, such as cistus or
sage, would suit it well.

*Erythronium
multiscapoideum*
Origin: California
Z: 5

❧ From the hills of the Sierra Nevada this little erythronium is a tough customer, despite its delicate appearance. Its flowers in April are creamy white of a silken texture with well-separated petals which twist as the flower ages. The centre of the flower is a striking lemon yellow. Several flowers are borne on each plant and they rise well above the foliage carried on slender brown stems. The flowers face outwards resembling diminutive narcissi. The leaves are narrow, undulating and finely mottled with brown markings. This is a plant for the woodland garden in a position of light shade where it will make one of the loveliest ornamental plantings you could have.

Erythronium revolutum
Origin: W. N. America
Z: 5

❧ In the wild this erythronium flourishes in moist coniferous woodland, its pale pink flowers glowing in the half-light. It is variable in colour, some plants have almost white flowers but others have a much deeper pink. The flowers are especially graceful, the buds drooping elegantly before unfurling and bending their petals backwards. The petals of those plants with deeper colouring show much darker veins along their

backs. The foliage is of the trout lily kind, undulating and strikingly marked with aqueous marbling, chocolate-maroon against almost jade green. In the garden it will flower in March or April and in conditions close to its natural habitat will naturalise easily. At Knightshayes Court in Devon a particularly good rich pink form, named 'Knightshayes Pink', has established a vast colony under old trees. These deeper pink clones are known collectively as the 'Johnsonii Group'. *E. revolutum* flourishes in the same conditions as bluebells (*Hyacinthoides non-scripta*) with which the dark pink kinds, flowering at the same time, will make a resplendent mixture.

Erythronium tuolumnense
Origin: California
Z: 5

❧ Among the first erythroniums in flower – often out in the middle of March in southern England – *E. tuolumnense* has a character all of its own. Its flower buds are green-gold on the outside and open into downward-hanging lemon-yellow trumpets which eventually open almost flat, like a little wild daffodil. They are smaller than other erythroniums and the petals never curve fully backwards as they do in most other kinds. Up to three flowers are carried on each stem which rises to 8in/12.5cm high. The foliage is pale green, with a shining surface, and an upwards habit of growth forming a distinguished background for the flowers. In the wild it grows on acid soil in woodland.

Illustration opposite:
Eucomis bicolor 'Alba'

Illustration:
Erythronium 'Pagoda'

In the garden it will do best in part-shade in humus-rich soil. Plant it with other shade-loving plants of striking foliage such as *Arum italicum* ssp. *italicum* 'Marmoratum'. *E.* 'Pagoda' is a hybrid with an unknown white-flowered species that is more vigorous than *E. tuolumnense*. Its stems rise higher – up to 15in/35cm – and it will produce several flowers on each one. The flowers are of a paler yellow and the petals are narrower and curve backwards in more characteristic erythronium fashion.

Eucomis

There are about 10 species of eucomis, all bulbs, in the family Hyacinthaceae/Liliaceae. They are known as pineapple lilies from the tuft of foliage at the top of the flowers which does indeed make them resemble pineapples. The name eucomis is derived from the Greek for 'beautiful hair.'

Eucomis bicolor
Origin: South Africa
Z: 8

❧ This marvellously decorative plant is rarely seen but is increasingly available in specialist nurseries. From a mound of gleaming, undulating strap-like leaves the fleshy, spotted stems, rising to 18in/45cm high, produce their exotic flowers in July. These consist of a bottle-brush of flowerlets, deep red and lime green,

crowned with a flamboyant tuft of lime-green leaves. Like most of the eucomis tribe they give off a horrible smell, not unlike rotting meat, and attractive to flies. The flowers last for weeks, well into September. It should be planted a good 6in/15cm deep in a well-protected position that gets plenty of sun. It needs soil that is both rich and well-drained. It makes an especially distinguished pot plant, either alone or as the dramatic centrepiece of a mixed planting. It may be propagated by seed or by dividing the bulbs in late autumn. *E. bicolor* 'Alba' is a cultivar with pale lime-green flowers and no spots on the flowering stem.

Eucomis comosa
Origin: South Africa
Z: 8

❧ This stately eucomis, previously known as *E. punctata*, throws out flowering stems that rise as high as 30in/75cm erupting from handsome gleaming foliage. The flowers, which open in July or August, are variable in colour from creamy pink to purple. One authority describes their scent as 'coconut ice, elderflower and jasmine' – a heady cocktail. It has an exceptionally long flowering season, often lasting well into October. Some forms have especially attractive leaves flushed with purple. As an architectural plant, of exquisite detail, this is one of the finest late-flowering bulbs, performing at a time when few other plants can equal its exotic beauty. If you can find the right place in a border – with plenty of sun and rich, moist soil – it will give tremendous pleasure and mix easily with other plants. Failing that, it is a marvellous plant for a large pot or for use as a most exotic bedding plant.

Fritillaria

The fritillaries are among the most irresistible of bulbous plants. There are about 100 species, in the family Liliaceae/Liliaceae, very widely distributed in the temperate regions of the western hemisphere. For owners of small gardens they make an admirable subject to collect – some of the trickier kinds may be grown in troughs or pots and given precisely the conditions they require.

Fritillaria acmopetala
Origin: Asia, Mediterranean
Z: 7

This fritillary is among the most elegant of its tribe. Its flowers appear in May, downward hanging bells whose petals curl back sharply at their tips. They are most strikingly coloured, alternate petals pale green and purple-brown. The flowers are carried at the tips of slender stems up to 10in/25cm high with wavy pale green leaves arranged alternately along the stem. In the wild it grows in clearings in woodland and in pastures. In the garden it will flourish in rich light soil in dappled shade. It would be well worth trying in a meadow garden where its relatively tall stems would give it great presence.

Fritillaria assyriaca
Origin: Turkey
Z: 8

The flowers of this fritillary are bell-shaped, hanging gracefully downwards on the tips of arched stems up to 12in/30cm high. They are deep maroon in colour, gradually fading into yellow towards the tips of the petals and the whole flower has a dusty sheen to it. The foliage is a fresh green, rising taller than the

flowering stems. In the wild it grows in poor, stony soil in exposed places as high as 8,000ft/2,500m. In the garden, where it will flower in March or April, it must have a sunny site and very well drained soil. Its colouring associates particularly well with pale grey foliage of plants such as the narrow-leafed *Salvia lavandulifolia* which needs similar growing conditions.

Fritillaria camschatcensis
Origin: N. E. Asia, W. North America, Japan
Z: 4

This is a beautiful fritillary whose flower colour varies considerably. The flowers open in May, grouped together at the tips of fleshy pale green stems which rise to about 12in/30cm. Up to six flowers are carried on each stem and they are downward hanging, rather stubby, bell-shaped with the petals opening out gracefully at their tips. They vary in colour from grey-purple, resembling slate, to a marvellous black-purple. They have bright yellow anthers. The leaves, ribbed and of a fresh green, are carried in whorls about the stem. It is a plant of woodland and meadows. In the garden it will do best in peaty soil in a position that is at least part-shaded.

Fritillaria imperialis
Origin: Asia
Z: 4

The crown imperial is the largest and most spectacular of the fritillaries – growing as tall as 5ft/1.5m. From the moment its fresh green growth erupts from the soil in February to its dramatic flowering some six weeks later it is one of most

Illustration:
Fritillaria imperialis

ornamental of garden plants. Even before its growth
emerges its presence is made known by its distinct foxy
smell which many gardeners hate. I find it irresistibly
attractive, one of the most evocative scents of spring.
The foliage is a gleaming green, with leaves twisting
decoratively about the fleshy purple-brown stem. The
flowers are held in a group of up to five disposed about
the stem, hanging downwards, and crowned by a
dashing tuft of pointed leaves. They are red or
orange-red, with striking veining in a deeper colour.
They never open out fully but remain tubular and
always pointing down. But it is worth raising a flower
to peer at the dazzling interior with crystal drops of
nectar held in suspension at the bases of petals and
curious hinged anthers. Equally beautiful in all respects
is the pale yellow-flowered cultivar, *F. imperialis*
'Maxima Lutea'. There are other curiosities such as one
with variegated foliage, *F. imperialis* 'Aureomarginata',

and one with two tiers of flowers, *F. imperialis* 'Prolifera'.

As the crown imperial is such an exceptional garden plant, and one which will settle down and naturalise in appropriate conditions, it is worth taking trouble to cultivate it well. It will flourish in rich soil in sun or semi-shade. I grow it underneath the deciduous ornamental tree *Amelanchier canadensis* whose pink-flushed buds are just breaking as the crown imperials' flowers appear. It must have rich, fertile soil that never dries out. The bulb should be planted no less than 6in/15cm deep. In heavy soil it is best to put a layer of coarse grit at the bottom of the hole. The bulb is very large and has an indentation at the top which, if planted upright, will allow water to gather and rot the bulb which should, therefore, be planted on its side. It will need feeding in later life – a thick mulch of compost before the leaves appear is beneficial. Foliar feeding after the flowers have faded will also build up strength.

Fritillaria meleagris
Origin: Europe
Z: 4

❧ The snake's head fritillary has curious lantern-shaped flowers in April. They hang downwards and are a rich maroon in colouring, attractively chequered. Each petal is ribbed in a deeper colour and tipped with gold. In any community of flowers some are white with greenish markings. They are held on stems up to 12in/30cm high, a pale glaucous grey with alternate bending leaves. In nature they are found in moist meadow land and in the garden they need a position in sun or light shade where the soil does not dry out in the summer. They are at their most beautiful as part of a mixed spring planting with bluebells (*Hyacinthoides non-scripta*), daffodils and wood anemone (*Anemone nemorosa*) in long grass. They will seed themselves and naturalise rapidly in conditions that suit them. The quickest method of building up a community is to gather seed and sow it in trays, transplanting the seedlings when large enough.

Fritillaria michailovskyi
Origin: Turkey
Z: 7

❧ This little fritillary, no more than 6in/15cm in height, has flowers of dazzling colouring. They hanging downwards, loosely bell-shaped, swaying on their stems. The upper part of the flower is very deep maroon, almost chocolate-coloured, with a dusty texture and the rims of the petals are rich yellow, forming a bold band at the mouth. A well-established colony gives off a delicious scent of honey. It needs a warm, sunny site and is best in well-drained soil – if your soil is heavy it is essential to add grit. In its native habitat it is found in mountainous regions up to 10,000ft/3,000m and flowers in the summer. In temperate lowland areas it will flower in the early spring and is best in a site that is hot and dry in the summer. A position in front of smaller shrubs, such as sages, will suit it well. The purple-leafed sage *Salvia officinalis* 'Purpurascens' makes a striking foil for the flowers.

Fritillaria pallidiflora
Origin: Central Asia
Z: 3

❧ This is one of the best fritillaries for the woodland garden, with marvellous presence and distinction in all its details. In its native habitat it is found in shady places in mountainous regions. It has handsome pale glaucous foliage, wide and pointed and elegantly curved. The flowers in April or May are carried well above the foliage, hanging in groups of three from the tips of stems which rise 12in/30cm high. Each flower is a hanging bell with shapely petals, a pale creamy-yellow and sometimes flushed with pink. Petals are strikingly marked with an intricate pattern of veins and the flowers give off a curious musky scent. It is at its best in a shady position in rich, peaty soil. It is a wonderful companion for wood anemones.

Fritillaria persica
Origin: Iran, Turkey
Z: 5

❧ The Persian fritillary is one of the most dramatically beautiful of the genus. It rises to a height of 4ft/1.2m and its stems are crowned in April by tall spires of hanging, bell-shaped flowers. They are dusty purple in colour, with lemon yellow anthers making a sharp

contrast. Although fairly tough (in the wild it
flourishes in mountainous country at heights of up to
8,oooft/2,5oom) in the garden it needs rich soil and a
position in full sun. The foliage is handsome, with
broad grey-green leaves emerging horizontally from
the stem. The cultivar *F. persica* 'Adiyaman' has much
darker maroon, almost black, flowers. It makes an
excellent plant for a pot.

Fritillaria pyrenaica
Origin: France, Spain
Z: 5

❧ This fritillary from the meadows of the Pyrenees is
like a splendidly distinguished cousin of the snake's
head fritillary, *Fritillaria meleagris*. It grows to a height
of 12in/3ocm and flowers in April or May. The flowers
are pale yellow strikingly marked with a purple-brown
chequerboard pattern on the backs of the petals. The
shape of the flower is distinctive with rather square
'shoulders' at the top and recurved tips to the petals.
The flowering stem is glaucous green with wispy leaves

arranged alternately up the stem. It is found in the wild as high as 7,000ft/2,000m and in the garden makes a superb ingredient for a cultivated meadow or in dappled shade at the edge of woodland. There is also a very beautiful almost pure yellow form.

Fritillaria raddeana
Origin: SW Central Asia
Z: 4

❧ This is like a smaller, much simplified version of *Fritillaria imperialis* with which it shares that curious, foxy smell. It grows to a height of 18in/45cm and bears its flowers in groups with a tuft of leaves rising above. The flowers are single, of the palest creamy-yellow, hanging downwards and spreading elegantly sideways when fully open. The leaves are a glistening, pale green, and twist decoratively. The fleshy stems are slightly flushed with purple. It is very hardy but flowers early in the year – in a mild season as early as February. Thus it may need some protection and may be best in the alpine house or bulb frame. I am experimenting with growing it in a very sunny border, backed by a stone wall, with a background of the glaucous foliage of *Euphorbia characias* ssp. *wulfenii*.

Illustration opposite:
Galanthus nivalis

Galanthus

There are about fourteen species of snowdrop, all native to Europe, in the family Amaryllidaceae/ Liliaceae, but many of these are very changeable and there are as many as 600 cultivars. Snowdrops have been taken up by collectors many of whom look out for minute variations – such as the markings on the petals – that make little difference to the garden-worthiness of the plant. All species prefer a moist soil, neutral or alkaline, and a position in partial shade – many will flourish in complete shade. All those described below are winter flowering, in temperate places in a mild winter they will flower in January. The species *G. reginae-olgae*, scarcely distinguishable to the gardener's eye from the common *G. nivalis*, flowers in the autumn when I think it looks completely out of season. To me these are the quintessential bulbs of the winter looking marvellous with other bulbous plants that flower at the same time such as aconites (*Eranthis hyemalis*) and *Cyclamen coum*, and with the decorative foliage of emerging pulmonarias, hardy geraniums and *Arum italicum* ssp. *italicum* 'Marmoratum'. Colonies of them under deciduous trees – which often provide the perfect habitat – are a superlative winter ornament. They may be propagated by division of clumps which is best done as the leaves begin to wither.

Illustration: *Galanthus elwesii* 'Flore Pleno'

Galanthus elwesii
Origin: The Balkans
Z: 6

❧ This Balkan snowdrop is like a much larger cousin of *G. nivalis*. Its foliage is broad and strap-like, up to 12in/30cm tall, and the flowers are more substantial. There is a good double form *G. elwesii* 'Flore Pleno'. This is a snowdrop of commanding presence, holding its own in decorative impact with ferns, hellebores and the emerging foliage of herbaceous plants.

Galanthus nivalis
Origin: Europe
Z: 4

❧ *Galanthus nivalis* is the common snowdrop native or naturalised in many parts of Europe. It is an irresistibly elegant and attractive little plant, one of the few plants that all gardeners will want to possess. The leaves are very slender and the flowers in January or February, up to 3/4in/2cm in length, are carried on stems 4–5in/10–12.5cm long. The flowers hang downwards, with petals slightly separate. A double form *G. nivalis* 'Flore Pleno' is also attractive. *G.* 'S. Arnott', of uncertain derivation but related to *G. nivalis*, is very vigorous, with more rounded flowers which have a strong honey scent.

Galtonia

There are four species of galtonia in the family Hyacinthaceae/Liliaceae. They are all bulbous and native to southern Africa.

Galtonia candicans
Origin: South Africa
Z: 5

❧ The summer hyacinth, as it sometimes called, is a prince among late summer-flowering bulbs. In late July or August it unfurls its splendid flowers. A tall, fleshy stem, glaucous green, up to 36in/90cm high, is crowned by a spire of hanging bell-like flowers, creamy white with a green base, looking like those of a giant and aristocratic cousin of the spring snowflake (*Leucojum*). The broad strap-like leaves with a lustrous surface form a handsome base to the stems. It must have rich, moist soil, in a sunny position, where it will settle down and seed. As a versatile bulb for the border it has few rivals in this season. It will go easily with almost any other planting but I have seen it in a Scottish garden looking magnificent planted in quantity among pale apricot coloured lilies. In cold gardens, it makes an excellent plant for a large container. It may be propagated by seed or by removing bulbils.

Galtonia princeps
Origin: South Africa
Z: 8

❧ This rather tender galtonia, smaller than *G. candicans* described above, is easy to overlook in the jumble of a border but the more it is studied the more irresistible it appears. Its flowering stem, up to 24in/60cm tall, bears a bold tuft of flower buds, like a giant plump ear of wheat without the whiskers. The individual flowers open in July, small pistachio-green trumpets, opening out at the tips with pointed petals. It must have rich, moist soil and a sunny position. Try and find a place where its subtle character will not be swamped by the more obvious charms of flashier plants. It may be propagated by seed or by removing bulbils. It looks wonderful rising above pale cream eschscholzias. *Galtonia viridiflora* has a similar colour of flowers, with a touch of yellow, which are held in looser, more elongated and taller spires.

Galtonia regalis
Origin: South Africa
Z: 8

❧ In many ways this rare galtonia is the finest of them all. In August it throws out 36in/90cm curving flower stems from which the flowers are suspended. It does not make such a well-defined flower-head as *G. candicans*, the flowers being more separate and well distributed along the stem. Each flower is a hanging bell, the colour of the flesh of a ripe avocado. It has the

most decorative foliage of all the galtonias. Strap-like, a good 3in/8cm broad, a beautiful lustrous lime green, the leaves curve hither and thither at the base of the flowering stems creating a lively effect. There is a sparkling freshness about the whole plant which makes one think more of the spring than the sultry days of late summer. It must have a sunny position in sharply drained soil and it will look superlative among the smaller shrubs that enjoy a similar position – cistus, lavenders, sages or santolina.

Geranium

Geranium tuberosum
Origin: Mediterranean
Z: 8

Most species of geranium, of which there about 300 in the family Geraniaceae, are herbaceous, or slightly woody, perennials. The one described here is unique in the genus because it is tuberous.

❧ This is among the first geraniums to flower, towards the beginning of May. Its flowers are a cheerful rich violet with silken petals crisply veined in a much deeper colour. It has very fine foliage with intricately cut

125

leaves which form striking hummocks about 8in/20cm high. It must have a sunny position and will flower well in quite poor soil. In the right position it will scatter seed liberally and form a self-perpetuating colony. It is an excellent plant for the front of a border intermingled perhaps with pinks whose pale grey foliage makes a good background for the sharp colour of the geranium's flowers.

Gladiolus

The modern florist's gladiolus is derived from hybrids of several South African species but these are not hardy in European gardens. However, there are some species, or cultivars close to them, that are hardy and make admirable garden plants. There are in all about 200 species, all corms, in the family Iridaceae.

Gladiolus callianthus
Origin: Tropical Africa
Z: 9

❧ Also known as *Acidanthera murielae* this very attractive gladiolus is tender but I have found that it is a very easy pot plant and the corms multiply obligingly with little trouble. It puts out stiff blade-like foliage above which the flowering stems rise to a height of about 36in/90cm. The flowers appear in August or September, dazzling white with spreading pointed petals and marked with a deep purple splodge in the throat. They are deliciously scented, with a hint of almonds. I have grown it in a large pot where it makes a marvellous ornament on a terrace. The corms are quite small and may be packed in 2in/5cm apart. The pots will need plenty of watering up to flowering time and, when the foliage has died away, the corms should be dried out and kept in a frost-free place. In spring they should be potted in new compost and may be started off in a cool greenhouse which will encourage early growth. In sheltered gardens it would be worth trying it in a very sunny position in well-drained soil, planting the corms a good 5in/12.5cm deep.

Gladiolus communis ssp. *byzantinus*
Origin: Mediterranean
Z: 6

❧ The Byzantine gladiolus is one of the most exotic plants that will naturalise easily – almost invasively – in temperate gardens. It has stiff blade-like leaves above which the flowering stems – 24in/60cm high – are tipped with groups of flower buds, flushed with bronze, that curve over like the beak of some strange bird. As the buds swell they take on a more pronounced purple colour with a decorative bloom, opening out at the end of May into splendid, outrageous magenta trumpets with white stripes in their throats. There are, in fact, two colours in the flowers – crimson and a purple which has an iridescent sheen; the two intermingle to appear as a single colour. It will seed itself benignly in the garden, scattering its dazzling colour in odd corners. It may also be propagated by detaching the cormlets that are formed at the base. It does well in poor, thin soil but must have a sunny position. In our garden it has dashing presence in a mixture of love-in-the-mist, *Nigella damascena*, and columbines, *Aquilegia communis,* which include the pink and white flowered 'Norah Barlow'. It is also an excellent plant for a purple and red border – rising above the sombre purple of the shrubby *Salvia purpurascens* or intermingled with the rose-pink *Cistus purpureus*. The emphatic foliage is always decorative, giving crisp architectural form in a border.

Gladiolus papilio
Origin: South Africa
Z: 8

❧ The smaller species gladiolus are plants whose attractions gardeners should explore more deeply. The leaves of *G. papilio* (also known as *G. purpureo-auratus*) are narrow, upright, of a striking pale glaucous grey. The flowers open in August, long pointed buds, rosy-purple smudged with lime-green, several borne on the wiry, arching stems. When the flowers open fully a yellow interior is visible. The flower stems rise

24in/60cm but are bowed down by the weight of the flowers which sway prettily on their slender stems. It should have a sunny position and it needs moisture to flower well – a fairly rich but well-drained soil suits it perfectly. In suitable conditions it will spread swiftly – in some lucky gardens becoming almost invasive. It is marvellous at the front of a border threading its way through low-growing shrubs. It is said to flower more profusely if it is constricted in a pot. I have seen it looking beautiful rising above a clump of *Zauschneria californica* which produces its vermilion flowers at the same time.

Gladiolus 'The Bride'
Origin: Garden
Z: 8

❧ This is among the most elegant of the smaller gladiolus. It produces its flowers in May or June, crisply white, shaped like an irregular wide-open trumpet with pointed petals. At the base of the petals the white merges into lime green giving the whole flower a sparklingly fresh appearance. The flower stems rise 24in/60cm, and are rather stiff and thus do not need staking. The corms should be planted 6in/15cm deep in well-drained but rich soil. A sunny position is essential. In mild years the leaves will often emerge in early spring and be vulnerable to frost. A light mulch of something like mushroom compost will protect them. It makes a superlative plant for pots. It is propagated by dividing the corms in the spring.

Hedychium

Hedychium coronarium
Origin: India
Z: 9

The ginger lilies are a genus of about 40 species in the family Zingiberaceae, rhizomatous perennials native to the tropics.

❧ This is one of the less tender hedychiums and makes a wonderfully ornamental exotic plant. The leaves are magnificent, 24in/60cm long, with a lustrous surface. The flowers in late summer or early autumn are white, gathered together in a bold spike, with a delicious, sweet scent. In its native habitat *H. coronarium* comes from very moist areas and may be grown standing in water. At the very least it will need rich, moist soil in the sunniest, most protected position you can find. Few gardeners have experimented with the hardiness of these magnificent species. They make marvellous plants for a large container or they may be treated as summer bedding plants where they are a superb ingredient for an exotic scheme. There are several available commercially. *H. densiflorum* (Zone 8) has orange flowers and *H. gardnerianum* (Zone 9) has yellow flowers and particularly handsome large leaves. They are easy to propagate by dividing the rhizomes in the spring.

Hemerocallis

Illustration opposite:
Hemerocallis 'Golden Chimes'

The daylilies are one of the most valuable, and easiest, of border plants. Their name comes from the Greek words for 'day' and 'beauty', a reference to the short life of each flower. There are about 15 species, all rhizomes, in the family Hemerocallidaceae/Liliaceae, all native to East Asia, China and Japan. Large numbers of cultivars have been bred in recent years, many from America. Over 20,000 cultivar names have been registered some of which are lovely garden plants but all too often the character of the whole plant has been sacrificed for gaudy and outlandish flowers which are sometimes of absurdly disproportionate size. Some of the names alone are enough to discourage the gardener – could you learn to love something called 'Little Fat Dazzler'? I describe in detail below some of the species. But the following modern cultivars are very

Illustration:
Hemerocallis fulva

good: 'American Revolution' is early flowering with
dark red flowers of velvet texture and elegantly tapered
form; 'Claudine' is rich red with a darker eye, a
marvellous ingredient of 'hot' schemes; 'Golden
Chimes', one of the oldest cultivars, is early flowering
with elegant small flowers, rich yellow within and
marked with golden-brown on the backs of the petals;
'Helle Berlineren' has striking pale pink-cream flowers
in June; 'Stafford' is for bold gardeners only – it has
dazzling scarlet flowers in June with a white rib.
Daylily cultivars come and go with great rapidity. As
with dahlias the best advice is to study the stock of a
specialist supplier and order colours and shapes to suit
your scheme. There are increasing numbers of dwarf
varieties which are both compact in height – as low as
12in/30cm – as well as having flowers in proportion,
sometimes as small as 2 1/2in/6cm in diameter. These
exist in a wide range of colours and are valuable for
small gardens.

All daylilies need moist, rich soil and will benefit
from foliar feeding from the moment their leaves

emerge in the spring until the flower buds are formed. A sunny position will produce the most floriferous plants. They may be propagated by dividing clumps in late summer or autumn. Species may be propagated by seed but with seed gathered in the garden there will usually be some loss of identity.

Hemerocallis citrina
Origin: China
Z: 4

❧ This daylily has sharp yellow flowers, deliciously scented. The flowers opening from June onwards last for several weeks and are up to 4in/10cm long, with well-separated petals which some garden authorities criticise for being too stiff. The petals are saved from too much rigidity by having attractively undulating edges. I think of them as being crisply defined and therefore having great presence in a mixed planting. It will make a compact clump, rising to 30in/75cm. Yellow is a difficult colour in the garden and this is one of the best. Graham Thomas called it the commuter's daylily because it did not open until the evening – a slight exaggeration, but it is certainly not an early riser.

Hemerocallis fulva 'Green Kwanso'
Origin: Japan
Z: 4

&❧ This exotic daylily was formerly known as *H.* 'Kwanso Flore Pleno'. Its flowers, which appear in July and flower for several weeks, are flamboyant both in colour and form. They are orange coloured and suffused with yellow, with a semi-double arrangement of petals. The outer petals curl backwards, thrusting the inner ones out. Some of the petals have smooth edges, others are frilled and undulating. There is nothing soothing about these flowers, they are full of excitement and vibrating with colour. They are carried at the tips of substantial stems that rise to a height of 36in/90cm. They come into their own in a bold colour scheme and I have seen them magnificently used in bold clumps among the deepest red dahlias and the soaring orange spotted flowers of *Lilium pardalinum* with crimson double-flowered nasturtiums lapping at their feet. The single *H. fulva*, one of the oldest daylilies in gardens, is a plant of great distinction with petals alternately coloured pale apricot and pale orange-brown.

Illustration opposite:
Hemerocallis fulva 'Green Kwanso'

Hemerocallis lilioasphodelus
Origin: Japan
Z: 4

&❧ This lovely daylily was formerly known as *H. flava*. It is among the earlier flowering kinds, producing its flowers in June. These are exquisitely shaped, wide open trumpets, like a very exotic daffodil. They are pale lemon yellow with a delicious, sweet scent. The stems, which are very slender, rise 30in/75cm tall. It will flower well in sun or in

part-shade – it is especially lovely in the dappled shade under a light canopy of leaves. The refinement of this daylily puts to shame so many coarse and overblown cultivars. It is a beautiful plant for the border, looking lovely with pale blue delphiniums.

Hemerocallis middendorfii
Origin: Asia
Z: 5

❧ This shows all the distinction of the best species daylilies. It flowers early in June, single lemon yellow trumpets with a marvellous scent. The petals curve backwards, making a flower of the greatest elegance. The plant is very upright with striking presence, the flowers borne on stiff stems up to 24in/60cm high among slender rush-like leaves. It is admirable with blues such as the silver-blue of *Geranium pratense* 'Mrs Kendall Clark' and the violet of *Viola cornuta*. It also goes well with the lime-green flowers of *Euphorbia characias* ssp. *wulfenii*.

Hermodactylus

There is only one species of hermodactylus, in the family Iridaceae.

Hermodactylus tuberosus
Origin: S. Europe
Z: 7

❧ Also known as the snake's head or widow iris, this lovely tuber is among the most decorative of all bulbous plants flowering in its season. The sweetly scented flowers open in April or May, iris-like, carried at the tips of stems 10in/25cm long. They are the colour of yellow Chartreuse but the falls are black, with the texture of velvet. In my own heavy clay, and plagued by slugs, it is a difficult plant. It is better in lighter, well-drained soil where it may seed itself prolifically. It should be planted in a sunny position or in part shade. In the garden it needs a simple setting, where it will not be overshadowed by plants of coarser charms. It is beautiful rising behind the pale green new foliage of *Alchemilla mollis*. It may be propagated by dividing clumps in autumn.

Hyacinthoides

There are four species, all annual bulbs, in the family Hyacinthaceae/Liliaceae, all native to Europe and N. Africa.

Hyacinthoides non-scripta
(syn. *Scilla non-scripta*)
Origin: Europe
Z: 5

❧ The common bluebell should only be planted in a naturalistic setting – in a formal bed it looks absurd. It is found in the wild in beech or oak woodland and anyone who has ever seen it flowering in April, spreading a smoky blue carpet through the trees, will

Illustration opposite:
Bluebells (*Hyacinthoides
non-scripta*) in an English
wood

recognise that nature had the best idea for its ideal position. The flowers are carried at the tips of fleshy stems, 12in/30cm tall, a mid blue with a hint of violet. Each flower is a diminutive bell, with petals curving back at the tip, hanging in clusters on one side of the stem. Occasional white forms are found and a much less desirable wishy-washy pink. In gardens it may cross with the undistinguished Spanish bluebell, *H. hispanica,* to produce prolific but dull offspring. The true bluebell must be planted on a grand scale – do not consider it if you have only a small garden. In woodland it will associate beautifully with substantial ornamental shrubs.

Ipheion

The name of this genus has been kicked about by botanists – it has been known under *Tristagma* and *Triteleia*. There are ten species of bulbous plants in the family Alliaceae/Liliaceae, all of which are native to South America.

Illustration: *Ipheion
uniflorum* 'Wisley Blue'

Ipheion uniflorum
Origin: Argentina,
Uruguay
Z: 6

❧ The spring starflower is one of those valuable small plants, very undemanding as to site, which will ornament all sorts of odd corners in the garden. It forms a hummock of fresh green strap-like leaves – with a powerful scent of onion. The flowers appear in February or March, borne at the tip of 6in/10cm stems,

Illustration: *Ipheion uniflorum* 'Rolf Fiedler'

each flower 1 1/2in/4cm across, star-shaped with pointed petals. They are pale blue with a deeper line down the centre of each petal, both front and back, and lemon-yellow stamens. The flowers have the faint but clear smell of honey, much magnified in the sun. In my own garden *Ipheion uniflorum* forms handsome clumps and it will flower well in semi-shade or in full sun, scattering itself obligingly. It looks beautiful with some of the pale yellows of other spring-flowering plants. I have seen it with the creamy-yellow double primrose *Primula vulgaris alba* 'Alba Plena'. It also makes a good plant for a pot, with its abundant foliage forming a fringe about the edge. It is very easily propagated by division of clumps. Some good cultivars are available, among them *I.uniflorum* 'Froyle Mill' with richer violet-blue flowers; *I. uniflorum* 'Wisley Blue', with strong blue flowers; and *I.uniflorum* 'Rolf Fiedler' with pale blue flowers with a striking white centre. *I. uniflorum album* is a good pure white form.

Iris

There are over 200 species of iris, in the family Iridaceae, widely distributed, but only in the temperate regions of the northern hemisphere. There are well over 800 cultivars commercially available in Britain, varying enormously in size and in colour – the name comes from Iris, the messenger in Greek

Illustration:
Pacific Coast iris

mythology who descended from the heavens in a
rainbow. They are all rhizomes or bulbs and may be
divided broadly into bearded and beardless kinds. The
bearded sorts, all of which are rhizomes, have a patch
of fuzz on the inside of the 'fall' – the downward-
curving petal which gives so many irises their especially
graceful character. It is from the species bearded irises
(such as the European native *I. variegata*) that the
majority of garden cultivars have been bred. There has
been a frenzy of iris breeding, with new varieties, often
of dubious identity, appearing – and disappearing –
with bewildering speed. In many of these the essential
gracefulness of the species has been lost in the pursuit
of new colours. The available colours range widely,
excluding only red and orange. Wild irises are found in
a wide range of habitats: moisture-loving species such
as *I. sibirica* or the European *I. pseudacorus*; those
flourishing in dry shade such as the valuable American
Pacific Coast irises (hybrids of *I. douglasiana* and other
species); and those relishing dry conditions such as the
North African *I. unguicularis* (formerly *I. stylosa*).
Many of these, and their garden varieties, provide

marvellous, easy garden plants, quite undemanding as to cultivation. Some of the most exquisite are often too small, or too demanding, to make adaptable garden plants. But many of them – such as *I. reticulata*, *I. bucharica* or *I. graebneriana* – make admirable plants for the pot or trough where they may be given exactly the conditions they need. Owners of small gardens, or those unable to tackle hard garden jobs, might well consider making a little collection of these; their cultivation is an absorbing subject and few plants can equal them for beauty. I have concentrated below on the garden-worthy species or those cultivars close in spirit to them.

Illustration: *Iris chrysographes* var. *rubella*

Iris chrysographes
Origin: Burma, China
Z: 7

❧ The distinctive quality of this iris is the exceptionally elegant form of the flowers which appear in June. The rich purple petals are long and narrow, and the falls bend sharply downwards to display a lively pattern of dark stripes on a yellow background. The upper petals curve upwards and the whole flower resembles a decorative and friendly insect. The flower stems rise to 24in/60cm and the upright leaves are almost as tall. It is best in moist soil in dappled shade in association with such things as Asiatic primulas and meconopsis (although it does not demand acid soil). *I. chrysographes* var. *rubella* (also known as *I. chrysographes* 'Rubra') is a particularly fine red-purple variety. Two outstanding cultivars have exceptionally dark, almost black, flowers: *I. chrysographes* 'Inshriach' and *I. chrysographes* 'Black Form'. All share the same handsome leaf and will form a statuesque clump of great structural presence.

Iris douglasiana
Origin: California, Oregon
Z: 7

&❧ The tribe of Pacific Coast irises has produced some splendidly decorative garden cultivars hybridising with other species such as *I. bracteata*, *I. innominata*, *I. munzii* and *I. tenax*. *I. douglasiana* has bold, sword-like leaves up to 30in/75cm long. The flowers are variable in colour but most have a darker or paler violet colour with darker veins and a yellow mark in the centre of the petals – but there is also a striking creamy-white form. The cultivars have produced some excellent colours including marvellous yellows such as 'Quintana'. Most of the Pacific Coast irises are found in acid sites at the edge of woodland. *I. douglasiana*, however, will tolerate alkaline soil and in my own garden, which is neutral, it flourishes. It has the additional advantage that it will flower well in shade; I have it growing under the evergreen *Photinia × fraseri* where it sparkles in the penumbra. It starts to flower quite early in April and continues producing flowers well into May. The rhizomes may be divided and replanted in enriched soil in early autumn.

Iris ensata
Origin: China, Japan
Z: 7

❧ Formerly known as *I. kaempferi*, the Japanese water-iris has in recent years given rise to a bewildering series of cultivars. The species, which flowers in June, has rich purple flowers carried on stems up to 36in/90cm tall. The buds before they open are almost black and the petals are etched with darker veins. It is a strikingly handsome flower. It needs very moist soil, at the margin of a stream or pool or planted in the shallows. It is at its best in part shade. Apart from the fairly straightforward white cultivar, *I. ensata* 'Alba', I do not know what to make of the frilly, pastel, parti-coloured, spotted, top-heavy cultivars with names like 'Prairie Love Song' or 'Moonlight Waves'. There are about 100 cultivars available commercially in Britain but only a handful are sold by more than one nursery. I cannot believe they will win the hearts of many gardeners in Britain.

Iris 'Katharine Hodgkin'
Origin: Garden
Z: 5

This cultivar was bred in the 1960s and has proved a
much admired plant. It arose as a hybrid of
I. histrioides and *I. winogradowii*. The flowers in
March, of extraordinary delicacy, are an exquisite
mixture of pale and deeper violet and rich yellow.
'Frank Elder' is similar in all respects but without the
deeper shade of purple. Although they are quite tough
in cultivation I think the best place for these treasures is
alone in a trough or in some corner of the garden where
their virtues may be appreciated in solitary splendour.
They are very easily propagated by potting up the
so-called 'rice-grain' bulbils which form at the base.

Illustration:
Iris laevigata 'Alba'

Iris laevigata
Origin: Asia
Z: 7

In its native habitat this iris is a plant of the
waterlogged banks of streams and pools. It may even
grow with its roots in the water. It will grow to
24in/60cm high and it flowers in June. The violet petals
spread widely creating a graceful winged shape which is
emphasised by the slender stripes of yellow which run
down the centre of each petal. It must have, at the very
least, moisture-retentive soil and it will flower well in
full sun or part shade. There are many good cultivars of
which the white variety, *I. laevigata* 'Alba' is especially
fine, with violet stripes on the inner petals. *I. laevigata*

is a refined plant and I have seen it well used in an underwater container in a formal stone-edged pool. It may be propagated by division in early autumn.

Iris latifolia
Origin: Spain
Z: 7

❧ Formerly known as *I. xiphioides*, this was most confusingly referred to as 'the English iris.' In its native habitat in the Spanish Pyrenees it is a plant of damp meadows. The flowers, which open in June, are a splendid rich violet-blue with petals of silky texture and falls marked with white and yellow. It grows up to 24in/60cm tall and the stems and leaves are a striking pale green with a hint of blue in it which makes a striking complement to the flowers. This is an iris which is suitable, as in nature, for naturalising in a meadow garden. It needs damp soil and an open sunny position. There is a good white form, *I. latifolia alba*, and a deeply disappointing pale violet form but neither has the *éclat* of the original.

Iris magnifica
Origin: Central Asia
Z: 5

❧ This is one of the Juno irises, all of which come from Central and Western Asia. It is, as the name says, a magnificent plant, one of the larger irises, with stems rising as high as 36in/90cm. Apart from the beauty of the flowers its great ornamental attraction is its statuesque form. From its thick fleshy stems handsome leaves stick out slightly upwards and in from their bases the flowering stems appear in late April or May. The flowers are white, of a silken texture with undulating edges. The falls of the petals are marked with yellow and in many plants there is a vague hint of blue towards the base of the petals. This is an easy garden plant, doing well in rich soil but demanding a sunny position. In its season few other herbaceous plants can compare for its dazzling impact. In the wild it is found in remote, dry mountainous places – as high

Illustration opposite:
Iris orientalis

as 5,000ft/1,500m – in the winter the rhizomes are buried deep under snow. In the garden it will prove a trouble-free plant, looking marvellous among the new foliage of astrantias, euphorbias and geraniums.

Iris missouriensis
Origin: W. United States
Z: 6

❦ It is the combination of foliage and flowers that makes the Missouri flag one of the finest species irises. The leaves are upright, quite narrow, blade-like, a very decorative pale glaucous grey. The flowers in May, carried on 24in/60cm stems, are pale lavender-blue, elegantly formed with narrow petals. The falls are veined with blue markings, as though painted with a fine brush, and marked with lemon yellow in the throat. In the wild it is found in alkaline soil and in the garden it should have a sunny position. In flower it is a dazzling sight – and the strongly architectural foliage continues to give pleasure long after flowering. There is a pretty white cultivar, *I. missouriensis* 'Alba'. *I. longipetala*, from the same area, is very similar.

Iris orientalis
Origin: Greece, Turkey
Z: 8

❦ Few irises can touch this for beauty and presence. In favourable conditions it will form a clump up to 5ft/1.5m tall with the flowers rising slightly above the tips of the leaves. The leaves are stiff, erect and blade-like, slightly glaucous in colour and looking marvellous with light shining through them. The flowers in June are white, with well-separated petals

which have decoratively crimped edges and whose centres are suffused with pale yellow. The falls arch downwards gracefully, and the other petals strain upwards. The flowers have a sweet, persistent scent. It is the mixture of boldness and delicacy that makes this an exceptional iris. In the wild it is a plant of wet places, often found in salt-marshes. In the garden it will need at the very least rich, moisture-retentive soil where it will soon form a statuesque clump. It will flower well in sun or part-shade and would be marvellous on the banks of a pool or stream under the light canopy of a tree.

Iris pallida
Origin: S.E. Europe
Z: 5

❧ The Dalmatian iris is one of the finest of the larger bearded irises. The large flowers in May are a beautiful clear lavender-blue with undulating petals intricately marked with darker veins. The beards are creamy white tipped with yellow. Flowering stems will rise to over 36in/90cm with as many as six flowers on each one. The leaves are a good pale glaucous green, quite short in relation to the flower stems, no more than 24in/60cm. There is an excellent variegated cultivar, *I. pallida* 'Variegata', whose leaves are broadly striped with creamy yellow along their length. This is a magnificent border iris. It is a large and characterful plant and any associated planting needs to be in scale; I have seen it looking marvellous with *Euphorbia characias* ssp. *wulfenii* 'Lambrook Gold' which has large pale lime-green flower head. Many hybrid cultivars among large bearded irises are derived from *I. pallida*, few rival it for beauty.

Iris pseudacorus
Origin: Europe
Z: 5

❧ The yellow flag, growing on the banks of streams and lakes, is a plant of bold character. It is an essential plant for any garden that has natural water. It has a statuesque form with stiffly upright blade-like leaves up to 4ft./1.2m high. They are glaucous in colour with finely shaded stripes running along their length. The flowers in May are a fine lemon yellow and their falls are marked with an etching of fine dark grey lines like a piece of Rococo embroidery. Several flowers are carried on each stem some of which are partly hidden among the foliage. There is a cultivar with handsomely variegated leaves, *I. pseudacorus* 'Variegata' with an especially good glaucous-grey colouring edged with silver. *I. pseudacorus* will form an emphatic clump that is quite able to hold its own with even the larger waterside plants like *Gunnera manicata*. It does not depend upon a waterside site – it will grow quite well in heavy, moisture-retentive soil. However, it looks its best in a naturalistic setting.

Illustration:
Iris sanguinea 'Alba'

Iris sanguinea
Origin: Japan, Korea, Russia
Z: 5

❧ This beautiful iris has much in common with *I. sibirica* but with a character all of its own. Mature clumps form striking sheaves of stiff, upright blade-like leaves which will rise as high as 36in/90cm having emphatic architectural presence. The flowers in late May or June are carried at the tips of slender stems and the flowers themselves are elegantly formed with narrow petals, the falls curving widely. They are rich purple in colour and the falls are marked with tiger stripes in yellow-brown. It is best in damp, rich soil and will flower very well in part shade. It is a marvellous woodland plant in a naturalistic setting, on the edge of a glade or stream. There is an exquisite white cultivar, *I. sanguinea* 'Alba'.

Iris sibirica
Origin: Europe, Russia,
Turkey
Z: 4

❧ The Siberian iris is a versatile plant – one of the easiest and most valuable of its tribe. The flowers, carried at the tips of very thin stems, appear in May. They are neatly formed, with narrow petals, an excellent rich purple-blue. The throats of the falls are intricately marked with a fretwork of dark purple, pale yellow and flecks of red-brown – exactly the same markings are to be seen on the outside of the buds before they open. The leaves are narrow, upright and slightly curving, rising 24in/60cm, almost as high as the flowers – they form a graceful sheaf. I grow it in several places in the garden and have found that it does equally well in sun or in part-shade. It prefers rich, heavy soil. I have seen it used as a lovely underplanting to the tree mallow *Abutilon × suntense* which has silver-lilac flowers which appear at the same time as the iris. It has been crossed with *I. sanguinea* to produce good cultivars. 'Heavenly Blue' has much of the upright *Sibirica* character with soft pale blue flowers. Among white-flowered cultivars 'White Queen' and 'White Swirl' are very attractive.

Illustration: *Iris unguicularis* 'Walter Butt'

Iris unguicularis
Origin: North Africa
Z: 8

❧ This, formerly known as *I. stylosa*, is one of the truly essential garden plants. The flowers are of marvellous beauty but the fact that they appear in winter, when nothing remotely as lovely is to be seen, is their trump card. The leaves give no hint of the beauties to come – they are grass-like, rather coarse, 24in/60cm long and some dead leaves are always apparent. The emergence of the flowers – at any time from late autumn to February – is one of the most exciting moments in the garden. Pale green fleshy stems, no more than 12in/30cm tall, unfurl revealing pale violet flowers. The petals curl back to show an interior striped with white and violet and marked with a smudge of lemon. The flowers are sweetly scented – all

Illustration: *Iris
unguicularis* 'Mary
Barnard'

the more noticeable if you bring them indoors and
display them in a vase. It must have a sunny, well
protected site and many gardeners say that it does best
in poor, stony soil. But in my own garden, with its rich
clay, it seems to flourish. I have it growing at the foot
of a bush of rosemary. The white form, *I. unguicularis
alba*, with the same lemon yellow markings is if
anything more beautiful. Particularly good cultivars are
I. unguicularis 'Walter Butt' with much paler colouring,
an almost silver violet; *I unguicularis* 'Mary Barnard'
with sparkling blue-purple flowers; and *I. unguicularis*
'Bob Thompson' with a rich purple colour. Many
authorities say that *I. unguicularis* resents disturbance.
I divided a plant of the white form in the late spring
and it flowered beautifully the following winter. The
one certain thing is that it will flower best if it has had a
good baking in the summer.

Ixiolirion

Ixiolirion tataricum
Origin: Asia
Z: 7

There are four species of ixiolirion in the family Amaryllidaceae/Liliaceae.

❧ This is a very beautiful summer-flowering bulb that needs hot, dry conditions to flower at its best. Its flowers in June or July are an intense violet-blue gathered together into loose sprays. The petals are very narrow, well separated and curve backwards. The leaves are thin, rather lax, glaucous green. It will grow to a height of about 12in/30cm. It is an excellent bulb to plant with low shrubs of Mediterranean character – artemisia, cistus and sage – which enjoy the same conditions and through which the ixiolirion may grow. *Ixiolirion tataricum* Ledebourii Group (syn. *I. ledebourii*) is similar but with more vivid violet flowers.

Kniphofia

Kniphofia caulescens
Origin: South Africa
Z: 7

Illustration opposite:
Kniphofia 'Royal Standard'

There are over 60 species of kniphofia, or red hot pokers, in the family Asphodelaceae/Liliaceae. They are all rhizomatous perennials and all native to the African continent. Most of those seen in gardens are cultivars of hybrids of impenetrably complicated origins. There are some very attractive plants among these, varying from the flamboyant 'Royal Standard' which will shoot up to 4ft/1.2m flaunting its vermilion and lemon flower heads, to the demure 'Snow Maiden', 24in/60cm tall, with white flowers. There are several excellent yellow cultivars. 'Sunningdale Yellow', 24in/60cm high, which produces good soft yellow flowers over a very long period, starting as early as late May – is a marvellous border plant. 'Percy's Pride' is a little taller in a sharper shade of yellow verging on lime green. Kniphofias may be propagated by dividing clumps in late autumn or in the spring.

❧ The flowers of this poker open in late summer, a refined salmon-pink which fades to soft cream. The flower-heads are broad at the base, narrowing towards the tip like a well-licked lollipop. The flowering stems are quite thick, rising to 4ft/1.2m and the foliage is a very decorative glaucous grey. It is marvellous rising

among pink phlox and *Tricyrtis formosana* which
flower at the same time. It is among the hardiest of the
pokers and is naturalised in certain parts of the east
coast of Scotland.

Kniphofia sparsa
Origin: South Africa
Z: 8

❧ The naming of this poker is uncertain – it may be a
synonym for *K. gracilis*. It is one of the smaller species,
no more than 30in/75cm tall, with exceptionally pretty
flowers which open in August. These are pink-brown –
exactly the colour of the gills of a fresh field mushroom
– fading to ivory at the base. As the individual flowers
fade long white stamens and orange anthers are thrust
out giving the lower part of the flower head a whiskery
appearance. The whole flower head is very narrow,
becoming pointed at the tip. The leaves curve about
elegantly at the foot of the flower stems. It would make
a beautiful plant for a pot and its cool elegance of form
and colour will make a striking contribution to the
late-summer border.

Kniphofia thomsonii var.
snowdenii
Origin: Central Africa
Z: 8

❧ This tender kniphofia is quite unlike others I describe here. It does not produce the usual bottle-brush inflorescence but instead its flowers hang outwards well spaced out along the stem which rises to a height of 36in/90cm. The flowers, which appear in July or August, are coral-pink, tubular and curve downwards resembling exotic miniature fruit. The leaves are grass-like, forming a sheaf about the base of the flowering stems. It needs a sunny position and in less favoured places a thick mulch will help to protect it in the winter. It may also be grown very successfully as a pot plant. I have seen it looking marvellous in a large urn with, at its base, the trailing *Convolvulus sabatius* with silver-blue flowers.

Kniphofia triangularis
Origin: South Africa
Z: 6

❧ The naming of this poker is a puzzle and it almost certainly covers more than one species. It is also known as *K. galpinii*. It has distinguished scarlet flowers with undertones of rust-orange carried on brown-tinged stems which rise to 36in/90cm. The flower heads are more elegant than other species with the individual flowers separated and showing their tubular shape. They open in late July or August and continue for a long period. The leaves are very thin, like grass. It gives admirable colour to the border.

Kniphofia uvaria 'Nobilis'
Origin: Garden
Z: 6

❧ This great poker is one of the most dramatic of all. Its flowering stems will rise to 6ft/1.8m bearing at their tips bold bottle-brushes of flowers a rich tomato red. The leaves are handsome, deep green, strap-like. It starts to flower in August but will continue more or less fortissimo well into autumn. Apart from the invigorating colour the larger kniphofias such as this have tremendous architectural presence, soaring above practically all other herbaceous planting. It is a superb ingredient for a richly coloured border. I have seen it marvellously used in front of a large bush of *Cotinus coggygria* 'Royal Purple' with the bold pleated leaves and scarlet flowers of *Crocosmia* 'Lucifer' at its feet.

Leucojum

A uthorities debate the meaning of the word leucojum; in Greek *leukos* means white but the second part of the word either means 'eye' or 'violet' (from the supposed scent) – neither of which is convincing. There are about ten species, all bulbous, in the family Amaryllidaceae/Liliaceae, native to North Africa, Europe and the Middle East.

Illustration opposite:
Leucojum vernum
'Gravetye Giant'

Leucojum aestivum
Origin: Europe
Z: 4

❧ The word *aestivum* means summer – misleading, because *L. aestivum* flowers in the spring. The snowflake is a showier version of the snowdrop, but with a character all of its own. It forms a generous clump with soaring blade-like leaves, a lustrous rich

green, which rise to 24in/60cm. The flowers appear in April, gathered in rows at the tips of stems which rise higher than the leaves. They hang like little white bells, with scalloped tips to the petals which are tipped with green. The cultivar *L. aestivum* 'Gravetye Giant' is much superior to the type, with a bolder presence. Its flowers are larger in relation to the leaves (which in the type are often so crowded as to obscure the flowers) and as many as five flowers are carried on each stem. It is at its best in a semi-shaded position in moist soil where it makes an admirable companion with some of the paler, more delicately formed, daffodils such as 'Thalia'. It makes a good plant for the woodland garden, with quite enough presence to embellish the foot of a flowering shrub or small tree. The bulbs should be planted fairly deep – no less than 5in/13cm. Propagate by dividing clumps after flowering.

Illustration: *Leucojum vernum* 'Carpaticum'

Leucojum vernum
Origin: Europe
Z: 5

᠅ The spring snowflake flowers very early, usually in February. It is a handsome plant with particularly attractive broad fresh green leaves, appearing well before the flowers, which arch gracefully, with the flower stems rising a little higher to 8in/20cm. The flowers hang downwards, white with smudges of green at the tips of the petals, resembling slightly squashed miniature lampshades. In the wild it is a plant of moist woodland and the dappled shade of hedgerows. In the garden, use it in more informal places – with ferns and hellebores, for example. The form *L. vernum* var. *carpathicum* has pretty yellow tips to the petals.

Libertia

There are about 20 species of libertia, all rhizomes, in the family Iridaceae native to South America and Australia.

Libertia formosa
Origin: Chile
Z: 8

❧ This handsome Chilean plant has a powerful architectural shape with upright blade-like evergreen foliage which rises to a height off 36in/90cm. The flowers, which appear from late May onwards, are arranged in generous, billowing upright umbels rising high above the foliage. Each flower has three rounded petals with white stamens and pale yellow anthers. The unopened flower buds are sheathed in bronze-brown making a striking contrast with the white of the opened flowers. It needs protection from the wind, is best in a sunny position and must have neutral to acid soil.

Illustration: *Lilium* 'Lady Bowes Lyon'

Lilium

There are about 100 species of lily, in the family Liliaceae/Liliaceae, among which are some of the most deservedly popular of garden plants. It is hard to imagine a garden without lilies. They are among the most spectacularly beautiful, and marvellously scented, of all hardy plants. Some have the reputation of being difficult or, like tulips, of flowering once only and gradually fading away. However, in certain conditions, there are many lilies that will settle down, reproduce and flower year in year out without the slightest difficulty. In my own garden, for example, I have two groups of the lovely panther lily, *Lilium pardalinum*, which is entirely self-supporting, producing its exquisite flowers without fail in early July. Even if it is impossible to provide suitable conditions for lilies in your garden beds, many may be grown most successfully in pots. Most lilies flower best in moisture-retentive soil that is neither too rich nor excessively acidic. For these reasons manure is not suitable and nor is bog peat. The best medium is natural

Illustration:
Lilium 'Enchantment'

leaf-mould. In the garden lily bulbs should be planted in a shady or part-shaded position and the soil should be enriched with leaf-mould. The perfect soil for the largest range of lilies is one that is slightly on the acid side of neutral. Slugs relish the bulbs and a handful or two of sharp sand surrounding each bulb will help to protect them. They should usually be planted 6in/15cm deep in late winter or early spring. This is no plant for miserly planting – bold clumps or drifts are best, naturalistically disposed. Once established, bulbs should not be disturbed, so be careful when digging nearby. Many lilies are surprisingly easy and quick to grow from seed, but some are very slow. *Lilium regale*, for example, often produces vast quantities of seed which, if sown in the spring, will have made a bulb by the end of the year and produce a flower the following summer. It is best, in fact, to remove this youthful flower so that energy is concentrated on forming a substantial bulb. If you sow seed every year you will soon establish a constant supply of flowering bulbs. The other technique for propagating lily bulbs is scaling, in which outer layers are removed from a mature bulb and placed in compost. The following year they will have formed bulbils at the base which will grow into mature, flowering bulbs. If you buy bulbs, by far the best, and cheapest, source is a reliable

Illustration opposite:
Lilium 'Stargazer'

wholesaler who specialises in them from whom you may order by the hundred and share them out with your friends. Non-specialist sources, particularly supermarkets, too often have badly stored and relatively expensive specimens.

The International Lily Register has divided lilies into 17 categories according to their genetic derivation or their general appearance ('flat star shaped flowers' and so forth). At any one time there are at least 200 cultivars commercially available but they come and go with bewildering speed and only a few are widely available over a long period of time (such as the purple spotted heavily scented 'Star Gazer'). Some really fine lily cultivars have a very short commercial life. The splendid orange-red 'Lady Bowes Lyon' was listed in the 1994/95 *The Plant Finder* as available from one source only. In the 1995/96 edition it is no longer available. The solution for the gardener is, if you want a particular colour or style of cultivar, scrutinise the suppliers' lists and order immediately. It is futile in a book, however, to recommend cultivars that quickly become unobtainable. There are countless hybrids, sometimes of wildly improbable form and colour. These may meet the need for a very specific colour for a certain effect but few are really good garden plants. The species described below will only become unavailable if they become extinct.

Lilium auratum
Origin: Japan
Z: 6

ﻢ This spectacular lily is one of the species that make one wonder why the breeders bother to hybridise new varieties. It is magnificently exotic, with the largest flowers of any lily. They appear in June, flaring wide open with petals well separated and curling back at the tips, as much as 10in/25cm across. The petals are white with smudges of yellow down the centre and a scattering of deep red freckles. The anthers are very striking, orange-brown, curved, thrust out at the tips of very long stamens. Lastly, the flowers suffuse the air with a deep, rich, languorous scent. It makes a big plant, as tall as 6ft/1.8m, that will certainly need

staking. In its native Japan it grows in volcanic ash and will not thrive on rich nourishment. Try and give it a warm place, with a certain amount of shade, with the best drainage possible in rather thin acid soil. It is wonderful in a very large container where you can give it the perfect conditions.

Lilium candidum
Origin: Greece
Z: 6

❧ The Madonna lily is a breathtaking sight in full flower in June or July. The flowering stems rise to 5ft/1.5m bearing up to six dazzling white trumpet flowers, up to 5in/12cm long, with orange-yellow anthers which emphasise the flowers' whiteness. The petals are slightly furrowed and the flowers give off a delicious fresh sweet scent. This, one of the oldest garden plants in the world, is famous for flourishing in neglected corners of cottage gardens and resisting the laborious attempts of gardeners to provide the conditions that will allow it to thrive. It is susceptible to viral disease and in cottage gardens it may be the only lily present and thus not exposed to alien infections. In the wild it grows in rocky, very dry places. The solution in the garden may be to plant it in a sunny, well-drained corner, far from other lilies. Unlike other lilies the bulbs should be planted quite shallowly, with the top of the bulb breaking the surface. Most Madonna lilies available are sterile, so the usual means of propagation is bulb scaling.

Lilium chalcedonicum
Origin: Greece
Z: 5

❧ The scarlet turk's cap lily is one of the earliest European lilies to have been used as a garden plant: it was known in the 17th century and a double form, now extinct, was seen in 18th-century gardens. It grows up to 5ft/1.5m high and produces its gleaming vermilion flowers with sharply reflexed petals in June. Each stem may carry as many as ten flowers. It is susceptible to viral infections but if planted in full sun and sharply drained soil, like *L. candidum* which comes from the same sort of habitat, it will do well. It is one of the finest of the red lilies and would make a superlative ingredient in a 'hot' coloured border. There is a very attractive old garden hybrid with *L. candidum*, *L.* × *testaceum*, with ivory petals and rich orange anthers.

Lilium longiflorum
Origin: Japan, Taiwan
Z: 8

❧ This superlative lily is scarcely reliably hardy, tending to succumb to cold, wet winter weather. However, it is among the easiest and quickest to raise from seed and, in pots, it makes a marvellous ornament for terrace or conservatory. It grows to about 36in/90cm producing its fabulous trumpet-shaped flowers in June. Before they open the buds are long, very pale green, full of exciting promise. The flowers are white, still with a hint of green, with the tips of the petals curving back only slightly. There is a dab of yellow in the throat, the anthers are tawny brown and the stigma lolls in the throat like a curiously shaped tongue. They give a deep, sweetly exotic scent which evokes hot tropical nights. They are the lilies most frequently seen in expensive flower shops and street markets and known as 'longies'. For the gardener its most precious use will be as a container plant that is exceptional in all respects.

Lilium martagon
Origin: Caucasus, Europe, Siberia
Z: 4

❧ The martagon lily is very variable and although some flower colours are more beautiful than others its stately presence and the way its flowers are carried is never less than beautiful. The flowering stems are up to 6ft/1.8m high with fresh green leaves borne in decorative whorls at regular intervals. The flowers are grouped in a loose raceme at the upper part, with as

Illustration: *Lilium
martagon* var. *album*

many as 40 flowers on each plant. The petals curve
backwards very sharply with the anthers thrown
outwards prominently, varying in colour from
egg-yolk yellow to orange-yellow. The stems bearing
each flower are horizontal with the flower carried at
the very tip hanging downwards like a miniature
lantern. In the wild it is usually found on alkaline soil
in woodland or scrub – as high as 8,000ft/2,400m. In
the garden it is one of the supreme woodland bulbs,
growing well in shade. Well-drained rich leaf mould
provides the perfect growing medium. To my eye it
never looks quite at home in a border, a wilder setting
seems much more appropriate. It varies in colour from
a rather wishy-washy pink-purple to a much finer rich
plum colour, and all have petals mottled with darker
spots. The white form *L. martagon* var. *album* is very
beautiful, as is the rich maroon *L. martagon* var.
cattaniae, neither of which has spotted flowers.
Propagation by seed or bulb scales is slow to produce
flowering bulbs but once established in an appropriate
site a community will flourish for years.

Lilium monadelphum
Origin: Caucasus
Z: 5

❧ From the upland meadows of the Caucasus, found
as high as 8,000ft/2,400m, this great lily is one of the
finest yellow-flowered kinds. It grows to a height of
about 5ft/1.4m, with rather stiff well-leafed stems, and
its flowers open in June. As many as 20 are held on
each stem, pale lemon yellow, with petals curving

Gardening with Bulbs

Illustration opposite:
Lilium monadelphum

sharply backwards. The unopened buds, hanging downwards, look like very exotic bananas, tipped top and bottom with a splash of gleaming crimson, still clearly visible in the fully open flower. A good clump in full flower is one of the most marvellous sights. In the garden it will grow well in semi-shade or full sun – I have seen it spectacularly naturalised on the edge of beech woods. It will flourish in different soils, heavy or light, acid or alkaline, but it must not be waterlogged. It may be propagated by seed or by bulb scales but bulbs take a long time to reach flowering size – up to five years. Once established they will settle down and flower without trouble for a very long time.

Lilium pardalinum
Origin: W. North America
Z: 5

❧ This is the lily for those who complain how difficult it is to establish lilies in the garden. I have two large communities of it, in rather different parts of the garden, one in full sun and the other in dappled shade. They flower every year with the greatest of ease – and what flowers! They open in July, the petals curving back upon themselves to make almost spherical shapes. The petals are tipped with orange-red but the bases are a warm yellow, splashed with spots of red-brown. The stamens are thrust well out below, tipped with orange-brown anthers held on the most delicate hinges, allowing them to flutter in the slightest breeze. The flowering stems rise 6ft/1.8m high, with several flowers carried at the top. The stems are ornamented with ruffs of leaves at regular intervals. They are quite stiff and, unlike other tall lilies, need little support except perhaps neighbouring plants to nudge against. It is easily propagated by dividing the rhizomatous bulbs which quickly form substantial clumps. In the garden it will do best in rich soil. I grow it rising above the daylily *Hemerocallis fulva* but it is also spectacular as an ingredient of a border of hot reds and purples. *L. pardalinum* var. *giganteum* is a form that will grow as tall as 8ft/2.5m with an even greater profusion of flowers, as many as 30 on a stem. The same plant is also known as *L. pardalinum* 'Red Sunset'.

Lilium pumilum
Origin: China, Korea, Mongolia, Siberia
Z: 5

ε❧ Among the best and easiest of the smaller-flowered lilies, *L. pumilum* has flowers of a dazzling colour. They are produced in late May or June, a lustrous vermilion with a gleaming surface. When fully open the petals curve right back upon themselves and orange stamens are thrust out. They have a curious smell, like chocolate. The flowering stems, which rise to about 24in/60cm high, are thin and wiry bearing flowers on slender side stems. In the wild it grows in quite open, exposed places but in the garden it will flower well in part shade where its gleaming, jewel-like flowers are displayed to great advantage. It is beautiful in a container. It produces huge quantities of seed but if you allow it to do so the bulbs will be deprived of nourishment. However, it is very easy to propagate by seed which will produce flowering plants within a year of sowing.

Lilium regale
Origin: China
Z: 5

ε❧ Despite all the hybridising of lilies, which has certainly produced some splendid colours, few lilies give such dependable pleasure as this marvellous Chinese species, introduced to the west by the great plant hunter E.H. Wilson in 1905. It flowers in June, producing white trumpets, up to 6in/15cm long, flushed with purple without and pale yellow within. The petals curve back gracefully at the tips and the flowers have a fabulous scent, sweet, spicy and intense.

A large bulb, growing vigorously in rich soil, will produce as many as 30 flowers on a single stem. It will flourish equally well in sun or in light shade and will tolerate an alkaline soil. It is one of the finest border plants and excellent in pots, suffusing a sitting area with its exquisite scent. It would come very high on many gardeners' lists of the absolutely essential plants. It may be propagated from seed sown in spring. Seedlings will produce flowers remarkably quickly, often in the second year after sowing, but these should be removed before any seed is set in order to concentrate all energy in the growth of the bulb.

Lilium speciosum var. *album*
Origin: Japan
Z: 6

❧ In all respects this is one of the most beautiful of the species lilies. It flowers in August, later than others, unfurling its wide open, almost flat, white trumpets and filling the air with its intense sweet fragrance. The backs of the petals are smudged with brown-pink and their edges undulate attractively. The long anthers are deep yellow. In full flower it gives the impression of prodigal abundance. It is rather prone to viral infections and, like *L. candidum* may be best grown on its own. It needs full sun and rich, deep soil. Once established it makes a wonderful border plant, rising to about 4ft/ 1.2m, but tending to flop with the great weight of flowers and needing support. It may be propagated by seed. There are several cultivars and the species itself varies, with more or less crimson flushing the petals. *L. speciosum* var. *roseum* is a fine pale pink and *L. speciosum* var. *rubrum* is carmine red.

Liriope

Liriope muscari
Origin: China, Japan
Z: 6

There are about five species if liriope in the family Convallariaceae/Liliaceae.

This autumn-flowering plant is valuable both for its striking flowers and ornamental foliage. The purple flowers which open in September or October are clustered together in upright spikes borne on a fleshy stem which rises to 12in/30cm. The leaves are blade-like and pointed, with a distinguished lustrous surface, and rise rather higher than the flowers. It must have a sunny position and rich, well-drained soil. It looks very beautiful with other bulbous plants which flower in the same period such as *Nerine bowdenii*. It may be propagated by dividing the rhizomes in the spring. There are several cultivars: *L. muscari* 'Monroe White' is pure white (also known as *L. muscari alba*) and *L. muscari* 'Gold-band' has gold variegated foliage.

Moraea

Moraea huttonii
Origin: South Africa
Z: 8

There are over 100 species of moraea, in the family Iridaceae, found only in southern Africa. The species described here is one of the very few hardy in European gardens.

This stately iris-like plant throws out bold flowering stems, 36in/90cm long, rather fleshy with flower buds sprouting alternately along the stem. The flowers which appear in May are a fine clear yellow with intricate dark grey markings on the petals which are well separated, giving them great elegance. They give off a sweet scent, light but quite distinct. The leaves are floppy but distinguished, each one up to 4ft/1.2m long and marked with dark veins along its whole length. The foliage forms a lovely mound from

which the flower stems emerge. This will need a very sunny position in light, well-drained soil. A large clump of it is a spectacular sight. It may be propagated by dividing the corms in the autumn.

Muscari

There are about 30 species, all bulbs, in the family Hyacinthaceae/Liliaceae. They are all quite small, not individually imposing, but forming decorative splashes of colour, sometimes richly scented, in spring.

Muscari armeniacum
Origin: Mediterranean
Z: 6

❧ The Armenian grape hyacinth has flowers of a wonderfully intense blue that seems to glow on even the dullest day. They are upright racemes composed of tightly-packed 'rugby-football-like bells' as the plant-hunter Reginald Farrer called them. The flowering head is up to 3in/8cm long held on a bright green stem of similar length. It is best in a sunny position and will flourish in poor soil provided it is well drained. It will make its greatest impact in the form of a bold clump – this is no plant for spotty planting. It looks especially beautifully with the pale silver grey of such plants as santolina or the slightly glaucous foliage of pinks – both of which enjoy the same conditions.

Muscari comosum
'Plumosum'
Origin: Garden
Z: 4

❧ The Mediterranean tassel hyacinth is a pretty plant, flowering in May or June, with racemes of flowers in upright spikes with the lower flowers brown/purple and the upper ones varying in colour from blue to violet and, best of all, an intense dark violet. It has the curious characteristic of some of the upper flowers being sterile and sprouting upwards at the top forming the stiff tassel that gives the plant its common name. In the form *M. comosum* 'Plumosum' *all* the flowers are sterile making the flower-head resemble a froth of flowers. The flower stem rises to 10in/25cm and as they are fairly slender they are often bent down by the weight of the flowers. It does best in full sun and is a highly decorative front-of-the-border plant.

Muscari macrocarpum
Origin: E. Mediterranean
Z: 7

❧ The flower stems rise from a mass of fleshy, glistening leaves up to 12in/30cm long. The flowers in April are borne at the tips of stems up to 6in/12.5cm long, upright racemes of diminutive tubes, like miniature uncurved bananas. Those at the top are a violet-brown colour, those at the bottom, banana yellow. The flowers give off a scent of tropical sweetness. It must have a sunny, warm position in well-drained soil – damp, heavy soil will kill it quickly. It is excellent at the front of a raised terrace, or in a rock garden, but in either case raised up so that its delicious perfume may be savoured. If your garden cannot provide the right conditions, it makes an excellent pot plant.

Illustration:
Narcissus tazetta

Narcissus

The name of the daffodil family comes from the
Greek for sleepiness or torpor, because of the
supposed narcotic properties of the bulbs. There are
about 50 species in the family Amaryllidaceae/
Liliaceae, native to southern Europe, the Mediterranean
region, China and Japan. Although there is only a small
number of species the number of named cultivars is
gigantic – running into several thousands. New
cultivars continue to appear but there seem to be no
valuable additions to the present repertory of truly
valuable garden plants. However, from the gardener's
point of view the choice is made easier because there
are only a few distinctive types.

For ease of classification daffodils have been divided
by the International Registration Authority into 12
groups defined either by the general appearance (for
example daffodils with emphatic 'trumpets' or coronas)
or by their origins (for example those descended from
species such as *N. tazetta* or *N. poeticus*). The species,

rarely seen in gardens, are an exceptionally attractive group but some of them have such specific cultivation requirements that make it difficult to accommodate them in the mixed plantings of most gardens. Most daffodil cultivars are undemanding, flourishing in soil of different kinds provided there is enough moisture and at the very least a position which gets some sunlight.

I think that the most beautiful use for daffodils is in a naturalistic context. A spreading group of the delicate *N. cyclamineus* under a deciduous tree or shrub, or of the deliciously scented poet's narcissus (*N. poeticus*) planted in the grass of an orchard, give greater pleasure than artful arrangements in the garden. Many of the

Illustration:
Narcissus cyclamineus

garden cultivars have a coarse shape – often with excessively large flowers of a strident, brassy colour. In massed plantings in public places the detail of an individual plant is much less important than the expanse of colour. In small private gardens, however, the defects of such plants will be all too visible. The most distinguished of the species, on the other hand, are always worth looking at in detail. Some of the more delicate of them are perhaps best planted in troughs or pots where not only is it possible to provide the perfect environment but they may be admired from close up. Several of the species have wonderful scent – very

Illustration: *Narcissus* 'Eystettensis'

rarely found in the highly hybridised varieties.

Most modern cultivars are derived from the western European wild daffodil, *N. pseudonarcissus,* forms of which have been valued garden plants for hundreds of years. The double-flowered *N.* 'Eystettensis' (also known as *N.* 'Capax Plenus') has been known since the 17th century and is still available commercially. It has very double flowers of an excellent pale creamy yellow.

Daffodils may be grown successfully in pots but it is only really worthwhile for the kinds that need special treatment. The scented species such as *N. jonquilla* or *N. tazetta* which I describe in more detail below make marvellous indoor ornaments or arranged on a sunny terrace. High temperatures will often prevent them from flowering. If you are proposing to force bulbs for early flowering they should be put in a protected but cool place (no more than 55°F). When the flower buds have been formed they may be brought indoors but they will still be best in the coolest part of the house.

Illustration: *Narcissus* 'Peeping Tom'

The soil should be slightly alkaline and moisture retentive; it should have no manure or extra nitrogen. As a general rule bulbs should be planted so that the depth of earth above them is equal to one-and-a-half bulb depths.

Probably the most common site for daffodils is in long grass on the verges of paths or in an orchard or meadow garden. Bulbs should be planted 6in/10cm deep, in a sunny or partly-shaded position. The grass should not be cut until the daffodil leaves have died down – which is usually at least six weeks after they have flowered. Although I have been successful with Pheasant's Eye narcissus, *N. poeticus*, in an orchard it is often the case that the vigorous hybrids are the most reliable for this purpose. At all events, do plant them in flowing drifts, avoiding any hint of regimentation. Bulbs should be planted no later than the end of September.

I have picked out a few of the species that I find particularly attractive – I give details of them below. So far as the cultivars are concerned, my taste is for those with paler colours and graceful petals. There are well over 700 named varieties on the market, an absurd proliferation. Here I pick out some that seem to me both to be especially attractive and to do well in gardens. They are arranged in order of season of

Illustration: *Narcissus*
'Jenny'

flowering. 'February Gold', 12in/30cm high, which
flowers in March rather than February, has a neat
trumpet of a cheerful yellow and elegantly-formed
paler petals that sweep backwards. 'February Silver' is a
much paler version. Both these are derived from *N.
cyclamineus*. 'Peeping Tom', also flowering in March,
has a similar character but is larger, up to 18in/45cm. It
has a long narrow trumpet with a crisply frilled mouth
and swept back petals. Its colour is a clear rich yellow –
striking and warm but without a hint of brassiness.
'Pride of Cornwall' flowers at the end of March. It is
derived from *N. tazetta* and has a deep gold flattened
trumpet and almost white petals. It grows up to
15in/35cm tall. 'Jenny', 10in/25cm high, has very pale
petals which curve backwards and a pale yellow
furrowed trumpet. It flowers in early April. 'Hawera'
which flowers in early April has several small flowers,
an excellent sprightly lemon yellow, which hang
downwards at the tips of quite short stems, no more
than 8in/15cm.

Illustration:
Narcissus bulbocodium

N. bulbocodium (Zone 6), the hoop petticoat daffodil, comes from the western Mediterranean region. The flowers which open in March have a distinctive shape, 2in/5cm long, strikingly resembling an old-fashioned full-length petticoat. They are held on short stems, 3–4in/8–10cm high, and the flowers are horizontal or droop slightly, varying in colour from pale lemon yellow to a rather richer gold. The leaves are very thin and grass-like. In the wild it grows in mountain meadows, usually on acidic soil and in the garden it is excellent for naturalising in short grass. The white hoop petticoat daffodil, *N. cantabricus*, is almost identical except with creamy white flowers and it is slightly less hardy (Zone 8).

N. cyclamineus (Zone 6), from Spain and Portugal has slender tubular flowers with the tepals swept straight back, which gives them a slight resemblance to the flowers of a cyclamen. The flowers are 2in/5cm long, warm yellow in colour, appearing in March carried on the tips of slender stems up to 9in/23cm tall, which allows them to nod in a breeze. In their native habitat they grow in damp meadows, often at the edge of a stream or pool. In the garden they naturalise well in partly shaded places under deciduous trees or shrubs, benefiting from dry conditions when dormant.

N. jonquilla (Zone 4) from Spain and Portugal has tall stems, 10in/25cm long, with several flowers held at the tip and grass-like foliage. The flowers appear in late March or April, rather flat and delicately formed, egg-yolk yellow, and giving off a marvellous rich, intense perfume. It must have a well-drained position in the sunniest part of the garden. In the wild it is found in rocky places in limestone. The recently discovered *N. cordubensis* is similar in all respects.

N. poeticus (Zone 4), the poet's, or pheasant's eye, narcissus is particularly valuable because it flowers very late – well into May – and has an exceptional rich, spicy scent. The flowers are white, with well-separated twisting petals framing a very short cupped corona which is lemon-yellow within and edged with a frilly scarlet rim. It is found in moist meadows, often in mountainous regions as high as 7,500ft/2,300m. In the garden it is marvellous planted in grass in a meadow garden or orchard, where it will naturalise easily. *N. poeticus recurvus*, sometimes known as Old Pheasant's Eye, has especially beautifully backwards sweeping petals. *N.* 'Actaea' is a Dutch cultivar with larger

Illustration:
Narcissus 'Actaea'

flowers, 3in/8cm across, and much taller stems – up to 18in/45cm tall – but with much of the beauty of the type and, best of all, the same fortissimo scent.

N. tazetta (Zone 8) is native to the Mediterranean region as far east as Iran. The flowers are carried in groups – as many as seven on one stem – at the tip of

Illustration opposite:
Narcissus 'Thalia'

tall stems up to 18in/45cm high. Each flower is up to 1 1/2in/4cm across and the cupped corona is lemon yellow framed by cream tepals which are wavy with pointed tips. The flowers are marvellously scented, intense, sweet and spicy – a bunch in a vase will perfume a room. With its soft colouring and sprays of delicate flowers at the tips of thin stems this is one of the most elegant of the daffodils. It is among the less hardy of those mentioned here and in many gardens will need a sunny, protected place. Here, flowering in March, it will provide distinctive elegance and scent the air on a warm day. The paper-white narcissus, *N. papyraceus*, is very similar in all respects except that the flowers are pure white. It is often used as a pot plant for forcing as a marvellously scented indoor ornament at Christmas time.

N. triandrus (Zone 4) is found in Portugal, Spain and North-western France in woods and open places. It produces its flowers in April, delicately formed, varying in colour from pale creamy-yellow to white, with petals swept gracefully back from the corona. It grows no more than 4in/10cm high. In the garden it will need light soil, good drainage and a sunny position. A modern cultivar with something of the species' character, but a tougher disposition, is *N.* 'Thalia', which is much larger, up to 12in/30cm high, with white flowers and a good scent.

Nectaroscordum

This genus has been separated from Allium with which it has very close connections. There are two species of bulbs in the family Alliaceae/Liliaceae.

Nectaroscordum siculum
Origin: E. Europe, Turkey
Z: 6

❧ This extremely decorative plant, formerly known as *Allium siculum*, has invasive tendencies but any keen gardener will be happy to put up with them for the beauty of the flowers. These, appearing in May and June, are gathered together in profuse umbels at the tips of stems that rise to 4ft/1.2m. Each flower is bell-shaped and hangs gracefully downwards on curving stems. There are two sub-species, *N. siculum*

ssp. *siculum* and *N. siculum* ssp. *bulgaricum* which have interbred, diluting their identity. The flowers of the former tend towards red-brown with shades of green whereas those of the latter are cream and pink. Plants deriving from these hybrids are widely found in gardens, showing varying character and colours. When the flowers fade the seed-pods point upwards, resembling rockets aimed at the sky. These seed-pods are strikingly ornamental but gardeners worried about too many offspring will remove them before they scatter their seeds. The beauty of this plant derives from its elegantly-shaped and subtly coloured flowers and its ability to rise up through densely packed plantings. The dark flowers of *N. siculum* ssp. *siculum* look wonderful with deep purple columbines, *Aquilegia vulgaris,* which flower at the same time.

Illustration:
Nerine bowdenii

Nerine

There are about 30 species of nerine, all bulbs and all native to South Africa, in the family Amaryllidaceae/Liliaceae. They are among the most beautiful of bulbous plants but virtually all the species, except *Nerine bowdenii* described below are winter flowering. However, some of the tender species are commercially available and make exquisite house plants – indeed far more beautiful than most of the plants normally thought of under that term.

Nerine bowdenii
Origin: South Africa
Z: 9

❧ As the last flowers disappear from the border, this nerine provides a wonderfully exotic flowering treat for the autumn. It flowers in September or October, the fleshy stems rising 24in/60cm crowned with a group of exquisitely formed sprightly pink trumpets. The petals have undulating margins and curve sharply backwards at the tips. They have blurred stripes of a deeper pink and purple anthers are thrust out. The foliage, appearing with the flowers, is strap-like, floppy and has a handsome glistening surface. It will remain in place until before flowering time the following year. It must be planted in a sunny place – the best groups I have seen have invariably been planted against a sunny wall. Although most authorities say that good drainage is also essential I have seen it very successfully grown in heavy clay. Bulbs should not be planted too deeply – the tips should be quite close to the surface. If groups become very congested they will not flower well and should be divided. A cultivar, *N. bowdenii* 'Mark Fenwick' (sometimes also known as *N. bowdenii* 'Fenwick's Variety'), is more floriferous and has larger flowers – by no means necessarily an improvement.

Nomocharis

There are about five species of nomocharis in the family Liliaceae/Liliaceae all native to Asia.

Nomocharis pardanthina punctulata
Origin: China
Z: 7

❧ This lily-like plant has irresistible charm. Its flowers in July are white, downward facing, generously freckled with purple spots and have a purple centre. The petals are slightly frilly and open out until the flower is almost completely flat, 3in/8cm across. The flowering stems rise as high 36in/90cm and as many as 20 flowers, hanging from horizontal stems, are borne on each stem. In the wild it enjoys high rainfall, acid soil and a generally cool climate. If you do not have such conditions it will not flourish. If you do, few bulbous plants will give more exquisite pleasure. It is, I think, very much a plant for a naturalistic setting of a woodland kind.

Ornithogalum

There are about 80 species of ornithogalum, all bulbs, in the family Hyacinthaceae/Liliaceae. They are native to Europe, Western Asia and southern Africa.

Ornithogalum arabicum
Origin: Mediterranean
Z: 9

❧ This very beautiful bulb is, alas, not very hardy but well worth trying if you can give it well-drained soil and a warm, sunny position. It flowers in May or June bearing a cluster of flowers at the tip of a fleshy stem which rises 18in/45cm high. Each flower is white, globe-shaped at first, and opens out to about 2in/5cm across revealing silky petals tipped with green, orange-yellow stamens and a striking glistening black ovary at the centre. As the plant ages the flowering stems tend to twist. In the rights conditions it will produce many bulbils providing a very easy means of propagation. *O. corymbosum* is said to be almost identical but hardier but I have never seen it listed by any nurseryman.

Ornithogalum montanum
Origin: Italy, W. Asia
Z: 6

❧ In the wild this charming little flower is found in mountainous, rocky regions, often in meadows. It is the sort of modest plant that is easy to overlook but closer inspection shows character and beauty. Its six-petalled star-shaped flowers appear in April or May. They are white and veins run the whole length of the petals giving them an elegant appearance. The backs of the petals are striped with green and unopened buds show green and white stripes. The prominent stamens are the palest yellow. It makes an attractive underplanting for ornamental shrubs and would look pretty in a meadow garden, provided the grass has been cut fairly short the previous season.

Ornithogalum nutans
Origin: Balkans
Z: 6

❧ The details of this little ornithogalum are very beautiful. The leaves are rather lax and the fleshy flower stem rises among them to a height of 12in/30cm. Up to a dozen little flowers are carried, appearing in April, hanging elegantly downwards on curved stems, each shaped like a wide trumpet with flaring petals. They are white but with grey-green stripes down the centre of the petals. The anthers are pale lemon yellow. The plant is exceptionally graceful in appearance with the flowers nodding in the slightest breeze. In the wild it grows in poor soil. In the garden it will do well in part shade but it must have good drainage. I have seen it naturalised in deciduous woodland and it is quite happy in completely dry soil in the summer. It will self seed lavishly or it may be propagated by dividing clumps of bulbs in the autumn.

Ornithogalum pyrenaicum
Origin: Europe
Z: 7

❧ This summer-flowering ornithogalum throws out tall, fleshy stems up to 36in/90cm. The flowers in June are gathered in tall racemes resembling heads of barley of exquisite elegance. Each flower has very narrow petals, pale green-gold with a green stripe down the centre. The foliage is strap-like, lax and forms a cushion at the base of the flowering stem. In the wild it is found in meadows and light woodland, as high as 4,000ft/1,200m. In the garden it is at its best in wilder places where it will bring a character of airy gracefulness. It would be superlative in a meadow garden – but you would have to leave grass-cutting very late to allow the seed to develop. I have seen a group of it grown successfully in a very large pot. Here, it should have rich moisture-retentive soil.

Oxalis

Oxalis oregana
Origin: W. North America
Z: 7

There are over 800 species of oxalis, in the family Oxalidaceae, very widely distributed.

❧ This American oxalis is a decorative plant for ground cover with the bonus of ornamental flowers. The foliage is boldly shaped, like clover with heart-shaped leaves. The flowers, which appear from late spring to mid summer vary: the best are an attractive pale pink, but they may be a rather wishy-washy mauve or white. In the wild it is a woodland plant, flourishing in moist leaf-mould. In the garden it is marvellous edging a path in a wilder part of the garden or providing a flowing pool of foliage under a shrub or tree. It will form a spreading colony and is easily propagated by division in the autumn.

Illustration:
Paradisea lusitanica

Paradisea

Paradisea liliastrum
Origin: S. Europe
Z: 7

There are only two species in this genus of rhizomatous plants belonging to the family Asphodelaceae/Liliaceae.

❧ This lovely plant is rarely seen in gardens, yet it presents no particular problems of cultivation. Fleshy stems rise to 24in/60cm with flowers grouped at the top. From waxy green-tipped buds the flowers open in June, slightly ragged trumpets with pointed petals, a

dazzling white with a glistening silken texture. The leaves are grouped together in a loose rosette about the base of the stem. It is not demanding as to soil but needs a sunny position. It is an admirable border plant, rising decoratively above lower closely-packed plants. It may be propagated by dividing the rhizomes in the autumn or, a much lengthier process, by seed. A larger-flowered cultivar, *P. liliastrum* 'Major', is occasionally seen but it lacks the delicacy of the type.

Paradisea lusitanica
Origin: Portugal, Spain
Z: 7

❧ In the right conditions this splendid plant will rise to 6ft/1.8m bearing its graceful plumes of flowers swaying elegantly on slender stems. The flower buds are gathered in a tightly-packed bunch at the tip of the stem. The flowers, which open in June, are white, star-shaped with long white stamens tipped with yellow. The foliage is inconspicuous in relation to the height of the flower stem, a ring of flaccid leaves about its base. It needs rich, moist soil and a position in full sun. Clumps of it are marvellous among shrub roses or rising among other plantings in a border. It is easy to propagate by division in autumn or by seed.

Polygonatum

The Greek origin of the name means 'many joints', referring to the angular form of the stems. There are about 30 species, all rhizomes, in the family Convallariaceae/Liliaceae, widely distributed in North America, Asia and Europe.

Polygonatum × hybridum
Origin: Garden
Z: 6

❧ The most commonly seen type of Solomon's Seal, an essential garden plant, is a hybrid between two species found in Asia and Europe, *P. multiflorum* and *P. odoratum*. The sweet scent of the latter has almost entirely disappeared in its hybrid but in other respects it is superior to its parents. The emerging shoots are very beautiful, a distinguished pale grey, and looking very edible. The stems, which retain a grey-purple base with a decorative bloom, grow as tall as 36in/90cm with leaves carried alternately along much of their length. These are beautifully shaped, furrowed, pale green becoming distinctly glaucous as they age. The stems flop over, sprawling horizontally, with the flowers hanging below, up to four on each stem. They are little white tubes no more than 1in/2.5cm long, opening out slightly at the mouth and tipped with green. I grow it in complete shade with ferns, the mottled *Arum italicum* ssp. *italicum* 'Marmoratum' and the ground carpeted with the shining leaves of *Asarum europaeum*. But it will also flourish in part shade and I have seen it looking wonderful in a white border as a backdrop to the tulip 'Spring Green' whose flowers are a soft white flushed with green. Two cultivars are fairly widely seen: *P. × hybridum* 'Flore Pleno' has double flowers but you will have to look very closely to notice; and one with a variegated leaf, *P. × hybridum* 'Striatum' (also known as 'Variegatum'), seems to have lost some of the hybrid vigour and always looks rather sickly to me. A much more beautiful variegated

Illustration opposite:
Polygonatum × *hybridum*

Solomon's Seal is *P. odoratum* var. *pluriflorum* 'Variegatum' whose leaves are most beautifully edged with cream, as though with brushstrokes by a skilful artist – a lovely effect. The flowers, furthermore, are sweetly scented. There is also a double-flowered cultivar, *P. odoratum* 'Flore Pleno', which has a blowsy charm of its own. All may be propagated by dividing clumps of rhizomes in the autumn.

Puschkinia

Puschkinia scilloides
Origin: Caucasus, Iran, Turkey
Z: 5

There is one species of puschkinia in the family Hyacinthaceae/Liliaceae.

ᚙ This, also known as *P. libanotica,* is a marvellous March-flowering bulb which has all the freshness of spring. Its flowers are carried on crowded spikes which rise to a height of 8in/20cm. Each flower is star-shaped, the palest blue with a crisp dark stripe down the middle of the petals. The leaves are a fresh glistening green, broad and pointed at the tip. It will flourish in almost any garden soil, in part shade as well as in full sun. In the wild the bulb dries out completely in the summer and in the garden it may enjoy similar conditions under deciduous shrubs such as magnolias or viburnums. *P. scilloides* var. *libanotica* is a form with slightly smaller flowers than the type. There is an attractive pure white cultivar of this form, *P. scilloides* var. *libanotica* 'Alba'.

Ranunculus

Ranunculus ficaria
Origin: Europe
Z: 5

The buttercup family, Ranunculaceae, contains about 400 herbaceous plants widely distributed throughout the temperate parts of the world. Most have fibrous roots but those described here are tuberous.

ᚙ The lesser celandine is a common European wild flower found in moist places in hedgerows, verges and deciduous woodland. The type, pretty though it may be in the wild, is too invasive for the more manicured parts of the garden. However there are several cultivars that have great garden presence. *R. ficaria* 'Brazen

Illustration: *Ranunculus ficaria* 'Salmon's White'

Hussy' has wonderful foliage, a very dark bronze-purple with deeply-veined heart-shaped leaves. The flowers, 1in/2.5cm across, are carried on fleshy stems and are a gleaming golden yellow, each flower lasting a short time, but appearing repeatedly in March. It will form a generous clump 4in/10cm high. *R. ficaria* 'Randall's White' has the green, marbled foliage of the type but with striking flowers with creamy-white petals and a bold tuft of lemon-yellow stamens. *R. ficaria* 'Salmon's White' is similar. *R. ficaria aurantiacus* (formerly *R. ficaria* 'Cupreus') has rich copper-coloured flowers. The curiously named *R. ficaria* 'Double Mud' is yellow with white tips to the petals and a dark centre. *R. ficaria flore-pleno* is a fully double form. *R. ficaria* is naturally very variable and new cultivars are constantly popping up. The foliage of all celandines dies down by the end of the spring and they may be propagated by dividing clumps.

Rhodohypoxis

There are about six species of rhodohypoxis in the family Hypoxidaceae, all perennials native to South Africa.

Rhodohypoxis baurii
Origin: South Africa
Z: 8

❧ This very decorative little tuber is often restricted to the rockery but, in the right conditions, it will make a good border plant. It is low growing, rising no more than 4in/10cm, forming a spreading cushion of narrow, hairy leaves. The flowers appear from June onwards and continue in great profusion for several weeks. They are star-shaped, with petals that overlap curiously at the centre. They range in colour from white to a rich cardinal red with some rather disappointing wishy-washy pinks and purples. In the wild they are found high in the Drakensberg mountains growing in well-drained peat. In the garden they will do best in a sunny exposure with plenty of moisture in the growing season – but none in the winter. They will not survive a winter in heavy, sodden soil. In the garden they find a decorative home at the front of a border intermingling with other plants of comparable scale which flower at the same time, such as diascias, trailing herbaceous potentillas and pinks. It is also excellent in pots – plant a single colour to the brim. It may be propagated by division. There are several cultivars of *R. baurii* and of hybrids. *R. baurii* 'Douglas' is a fine rich red and *R. baurii* 'Albrighton' is rosy red. *R. baurii* 'Margaret Rose' is a refined, pale pink. *R. baurii* 'Apple Blossom' is a very decorative mixture of green and pink.

Romulea

There are about 80 species of romulea, all corms, in the family Iridaceae and native to South Africa, Europe and the Mediterranean region. Many need protection in cooler European gardens but it is a delightful genus and would make an excellent subject to collect for the cool greenhouse.

Romulea nivalis
Origin: Lebanon
Z: 8

❧ The flowers are borne on wiry stems that rise to a height of 4in/10cm. They open in March or April, beautiful pale lilac trumpets with a vivid yellow base and dark smudging on the backs of the petals which are pointed and curl backwards gracefully. The leaves strongly resemble those of crocus – narrow, upright and pointed with a pale stripe running their length, rising rather higher than the flowers. In the wild it is found as high as 7,000ft/2,000m in places which are bone dry in summer. In the garden it should have gritty, well-drained soil in a sunny position.

Roscoea

There are about 15 species of roscoea in the ginger family, Zingiberaceae.

Roscoea auriculata
Origin: Himalaya
Z: 6

❧ Many people on first seeing this lovely plant assume it to be an orchid. It stirs into life and flowers late in the season, from August to September, bringing a note of spring-like freshness to the garden. The flowers are rich purple with petals like rather floppy wings, several flowers emerging one after the other, as a magician draws streamers from his hand. The flowering stems are up to 18in/45cm high, pale green and fleshy, with handsome broad shining leaves emerging horizontally. It will flower well in sun or in part shade and needs well-drained acid loam. It is a graceful and delicate plant, perhaps at its best in a little corner of its own.

Roscoea cautleoides
Origin: Himalayas
Z: 6

This beautiful plant looks at first sight like a slightly lopsided iris to which it is not even vaguely related. It flowers in June, unfurling the palest yellow petals which glisten like fine silk. Some petals flop downwards and have ruffled edges. The fleshy stems will grow to a height of 18in/45cm with attractive fresh green undulating leaves emerging horizontally up the stem. It is at its best in deep, rich acid soil with good drainage and need at least part shade. It may be propagated by dividing the rhizomes in early spring, or by seed. Many gardeners report self-sown seedlings. This is a lovely plant to follow after erythroniums, hellebores and snowdrops as it enjoys the same conditions. The rhizomes should be planted quite deep, with 6in/15cm of soil above their crowns.

Sanguinaria

Sanguinaria canadensis
Origin: E. N. America
Z: 4

The bloodroot, in the family Papaveraceae, has a single species. It is a rhizome whose name derives from its red sap.

The bloodroot is one of the most mysterious and beautiful of plants. Its shoots erupt from the earth in late March or April, pale, grey and fleshy. Flowers and foliage appear simultaneously, pale pinkish-white buds thrusting from the centre of rolled leaves. The flowers,

Illustration: *Sanguinaria
canadensis* 'Plena'

up to 2in/5cm across, borne on pink-tinged stems, are
the purest white, with pointed well separated petals and
upright lemon yellow stamens. The leaves –
wonderfully ornamental in their own right – are
strikingly lobed like those of a miniature fig, glaucous
green and deeply veined. In its native habitat bloodroot
ranges widely from Florida in the south to Canada in
the north and it flourishes in very different soils from
the slightly alkaline to the strongly acid. It is a
woodland plant and in the garden it must have at the
very least part shade. It demands moist, rich soil with
plenty of leafmould if possible. Here it will make an
excellent companion for other woodland plants – with
choice ferns, Solomon's Seal and trilliums. Even more
beautiful than the type is a cultivar with fully double
flowers, *S. canadensis* 'Plena', which is among the most
exquisite of all woodland plants. A rare pink cultivar,
S. canadensis 'Roseum', has the palest pink single
flowers and its foliage is flushed with a dusty mauve.
The rhizomes of bloodroot should be planted a good
4in/10cm deep. They are best planted in the autumn
which is also the season for dividing over-crowded
clumps. They may also be propagated by seed.

Schizostylis

There is just one species of schizostylis, a rhizome in the family Iridaceae.

Schizostylis coccinea
Origin: Southern Africa
Z: 6

❧ The Kaffir Lily, producing its cheerful flowers in late summer or early autumn is one of the most unexpected of bulbous plants. The single flowers are carried at the tips of thin wire-like stems up to 18in/45cm high. The flowers are red, varying in intensity and sometimes with a hint of orange. The petals are veined in a deeper colour and the stamens are unusually long and thin. In the wild it is found in damp places, often on the banks of streams. In the garden it will flower best in rich, moist soil in a sunny position. It is wonderful with grey-leafed plants, such as artemisias which late in the summer start to become blowsy and sprawling making a good background its sprightly colour. It is easy to propagate by dividing the rhizomes in the spring. There is a good white form, *S. coccinea alba*, and many cultivars: *S. coccinea* 'Jennifer' has good pale pink flowers with petals that are wider than the type; *S. coccinea* 'Major' has larger flowers which are rich red.

Scilla

There are about 90 species of scillas, all bulbs, in the family Hyacinthaceae/Liliaceae.

Scilla liliohyacinthus
Origin: France, Spain
Z: 6

❧ The Pyrenean squill combines decorative foliage with flowers of distinctive colour. They appear simultaneously in late April or May – the leaves are broad, strap-like, mid-green with a lustrous surface.

The flowers rise above, held in little umbels, each flower a diminutive pale violet star with deep purple anthers. In the wild it is found in meadows and woods – often as high as 7,000ft/2,000m. In the garden it is very versatile. In would look lovely in a meadow garden or orchard provided the grass has been kept fairly short – the flowering stems rise no more than 4in/10cm. The right sort of place would be where the grass has been cut late in the previous season, for example to display *Cyclamen coum* and *Anemone blanda*. There is an attractive rare, white cultivar, *S. liliohyacinthus* 'Album'.

Scilla messeniaca
Origin: Greece
Z: 8

❧ The Greek squill comes from the Kalamata region, famous for producing the finest olives, and it does indeed look very beautiful growing beneath the silver leaves of olive trees. The flowers in March are borne in little racemes, a beautiful grey-blue, carried on stems that rise to 4in/10cm high. The leaves are quite broad, strap-like, with a gleaming surface. This is a plant for naturalising so that it forms a spreading carpet below other plants. It looks beautiful with the cream-green of the flowers of the Corsican hellebore, *Helleborus argutifolius*. In the wild it likes shady places and is often found growing in pastures. It makes a

distinguished underplanting to such spring flowering
shrubs as *Corylopsis pauciflora* with whose soft yellow
flowers it goes admirably. Some of the smaller, simpler
narcissi with creamy yellow flowers also make
admirable partners.

Scilla mischtschenkoana
(syn. *S. tubergeniana*)
Origin: Iran, Russia
Z: 6

The first sign of this little bulb is the emerging
flower heads – a vigorous froth of blossom – bursting
from the soil in February. The flowers are either the
palest blue or white, with deep blue veins down the
back of each, carried in loose racemes, rising to
3–4in/8–10cm high. They are closely followed by
shining, strap-like leaves, to 6in/15cm long. It is best in
rich, moist soil, and in order that it should flower as
early as possible, plant it in a sunny position. It seeds
itself, and clumps may be divided, and it makes an
excellent plant for naturalising at the feet of deciduous
shrubs such as magnolias which will provide dry
summer conditions for its period of dormancy.

Scilla peruviana
Origin: W. Mediterranean
Z: 8

Despite its name this splendid bulb has nothing to
do with Peru. It flowers in May or June, producing
bold clusters of up to fifty flowers forming a rounded
cushion. The buds are a rich, deep purple before they
open into violet-blue star-shaped flowers with brilliant

yellow anthers at the centre. The flower heads are carried on fleshy stems, up to 10in/25cm tall, which rise above prolific shining leaves. It needs a sunny position and looks beautiful planted with smaller silver-leafed shrubs such as santolina (*Santolina chamaecyparissus*). There are paler coloured forms, including some that are an almost grey-violet, and a dazzling white cultivar, *S. peruviana* 'Alba'.

Scilla siberica
Origin: Armenia, Georgia, Turkey
Z: 5

≥ This is variable in colour and the best clones have flowers of a rich vivid blue with a hint of violet. The leaves are fleshy with a lustrous surface and the flowers, which open in March, are held at the tips of stems up to 4in/10cm high. The flowers are borne singly, hanging downwards like slightly crumpled lamp-shades, marked on the backs of their petals with a dark stripe. It adds dabs of brilliant colour to the early spring scene, mingling well with other ornamental planting. In my garden it grows in a narrow west-facing bed among the silvery patterned leaves of *Cyclamen hederifolium*. It is at its best with plants of similar small size, such as *Anemone blanda* and *Crocus tommasinianus* with whose grey-violet it looks lovely.

Sisyrinchium

There are about 90 species of sisyrinchium in the family Iridaceae widely distributed in America and Australasia.

Sisyrinchium striatum
Origin: S. America
Z: 8

&❧ It is easy to overlook the great merits of this common plant. Its stiff blade-like leaves are evergreen and very ornamental, grey-green in colour and rising to 18in/45cm, providing crisp architectural shape in the blur of a modest border. The flowers in June are an excellent colour, the softest creamy-yellow which harmonises easily with other colours. They are carried in profuse groups on erect stems, each flower like a widely open trumpet, emerging from curiously striped buds. *S. striatum* 'Aunt May' (sometimes wrongly called *S. striatum variegatum*) has especially distinguished leaves striped with swathes of creamy yellow. It is short lived but easily propagated by division in the autumn A sisyrinchium of unknown origins is the cultivar with the very odd name, *Sisyrinchium* 'Quaint and Queer' which has flowers that are cream and ochre, a charming mixture.

Smilacina

There are 25 species of smilacina, all rhizomes, in the family Convallariaceae/Liliaceae, native to Asia and America. They are sometimes called False Solomon's Seal but the species I describe below is entirely distinctive.

Smilacina racemosa
Origin: Central and North America
Z: 4

Among woodland plants this is one of the true aristocrats. Its common name is False Spikenard – a reference to its aromatic qualities which were thought to make it a substitute for the exotic true spikenard, an Indian plant used to make a costly ointment. The leaves are strikingly ornamental, 6in/15cm long, undulating and furrowed, a fresh pale green in colour. Their stems

Illustration:
Smilacina stellata

tend to flop sideways making the foliage resemble a lively sea of leaves. The flowers which appear at the tips of the stems in May are airy plumes, creamy white in colour tinged with green, and giving off a delicious sweet scent. In warm climates red berries are formed. This is a superb plant for a shady place, especially among large shrubs or trees in a woodland garden. It will spread to form an emphatic clump, rising 36in/90cm high. It needs a cool, shady position in lime-free soil. It may be propagated by division in the autumn. It has a much smaller cousin, *S. stellata* (syn. *Maianthemum stellata*), which has charm but spreads like wildfire so should be admitted only to the wildest corner of the garden.

Sternbergia

There are about seven species of sternbergia in the family Amaryllidaceae. They are sometimes referred to as 'autumn daffodil' which is not helpful as, although they are related, they do not in the slightest resemble daffodils.

Sternbergia candida
Origin: Turkey
Z: 7

❧ This spring-flowering sternbergia was discovered only twenty years ago and it was recently in the news because it has been over collected in the wild. However, it is now available commercially and well worth obtaining. The white flowers are carried at the tips of 6in/15cm high flowering stems which rise slightly higher than the grey-green strap-like foliage which appears at the same time. The flowers resemble those of a crocus but in full sun open fully and the petals separate giving the flower the appearance of an irregular star. In the wild it is found in dry, rocky places and on the edge of woodland. In the garden it should have a sunny site and good drainage.

Sternbergia lutea
Origin: Southern Europe
Z: 7

❧ There are very few plants in my garden that give more pleasure. In September or October the flowers erupt from the ground, luminous pale golden yellow, resembling a well-fed crocus, and rising 6in/15cm, looking wonderful among the first autumn leaves. The flowers never open out fully, always remaining slightly cupped even in bright sunshine. When the sun is not shining they are smoothly goblet-shaped. The leaves appear at the same time, lustrous green and strap-like, rising higher than the flowers. The foliage remains throughout the winter, dying in the following spring. It is said to flower best where the bulbs are constricted. Although some authorities say that it needs good drainage my plants flourish in our heavy clay, growing between old paving stones where they seem to relish the moisture. *S. lutea* ssp. *sicula* is very similar but more delicately formed as is *S. colchiciflora* which has more separate and pointed petals.

Tecophilaea

There are two species of *Tecophilaea* in the family Tecophilaeaceae/Liliaceae, both corms. They are native to the high Andes in Chile, found as high as 10,000ft/3,000m.

Tecophilaea cyanocrocus
Origin: Chile
Z: 9

❧ The Chilean crocus, probably extinct in the wild, is creeping back into cultivation through the efforts of conservation botanists. There are plans to reintroduce communities to the wild and it is now available through a few nurseries in Europe and the U.S.A. It is one of the most memorable of all small bulbs. The flowers, which appear in February or March, are of the most intense blue, equalled only by the blue of gentians. The flowers, up to 1 1/2in/3cm long, are loosely trumpet-shaped at first but soon open out fully and a paler throat is revealed, with veins running along the inside of the petals. The leaves are pale green and rise higher than the flowers which are half concealed by the foliage. In its native habitat a carpet of snow protects the Chilean crocus from the harsh winter conditions and it flowers in late autumn. In gardens with a temperate climate the best results will be obtained in a very sunny site in areas of low rainfall where it will flower in late winter or early spring – February to March. It has been successfully grown out of doors in gardens in the south of England in on the east coast of Ireland. It is a plant of such beauty that in less favourable climates it would be worth growing in a cold frame, or Alpine house, for the exhilaration of its dazzling colour. There are two cultivars: *T. cyanocrocus* 'Leichtlinii' with paler blue flowers, not as exciting as the type; and *T. cyanocrocus.* 'Violacea' with deep violet flowers – a good colour but without the piercing intensity of the type.

Tradescantia

There are about 70 species of tradescantia in the family Commelinaceae all native to America. Their name comes from the English royal gardener John Tradescant the Younger who made plant-hunting expeditions to Virginia in the mid 17th century, introducing several North American plants to English gardens.

Tradescantia virginiana
Origin: Eastern U.S.A.
Z: 7

❧ The spider lily is a long-flowering rhizome with striking foliage. The stems bearing leaves and flowers are jointed rather like a reed with leaves sticking out more or less horizontally. These are up to 10in/25cm long, narrow, pointed and with a pleat down the middle. Bunches of flower buds appear cupped between two leaves at the tips of the stems. The flowers which open in May are three-petalled, rounded, and varying in colour – white, blue or purple. In my own garden I have a clone that is white faintly flushed with pale blue. The yellow anthers are suspended above curious hairy sepals as fine as swan's down. It will form a bold upright plant up to 24in/60cm tall with the flowers half-concealed among the leaves. It is best in rich, heavy soil and will flower equally well in sun or part shade. It is easy to propagate by dividing clumps in the autumn. There is a group of cultivars of the hybrid *T. × andersoniana* which include 'J.C. Weguelin' with clear lavender flowers and a good white-flowered one, 'Osprey'. All these are valuable border plants

Tricyrtis

There are about 15 species of tricyrtis, or toad-lilies as they are called, all rhizomatous perennials, in the family Convallariaceae/Liliaceae.

Tricyrtis formosana
Origin: Taiwan
Z: 7

❧ This very decorative perennial is a valuable plant for the late summer border. It flowers in August or September but long before that its distinctive foliage has made its presence felt. The leaves are pointed, rounded, furrowed and their bases are transfixed by the stiff flower stem which rises to a height of 36in/90cm. The flowers are wonderfully exotic. The outer petals are narrow and pointed, white but scattered with spots of deep red. The inner petals have the same colouring but are arranged like a miniature palm-tree or gushing water jet in the centre of the flower which is marked with yellow at its base. The stems and flower-buds are covered with fine hairs. It grows and flowers well in semi-shade and is best in rich, moist soil. I have it mixed, accidentally, with some rather shrill border phlox, pink and carmine, with which it makes a cheerful mixture. It is easy to propagate by dividing the rhizomes in the spring. A separate group, known as Stolonifera Group, has more lax growth, softer foliage and more spots of colour. A cultivar of unknown origin, *T.* 'White Towers' has white flowers and grey-green foliage. *T.* 'Tojen' has solid pink flowers.

Tricyrtis hirta
Origin: Japan
Z: 5

❧ This Japanese toad lily has flowering stems that rise to 36in/90cm with flowers that are similarly elaborate as those of *T. formosana* but they are borne in the leaf axils and the petals are more upright. They are white and have maroon spots. The foliage is striking, boldly heart-shaped and sticking out horizontally from the stem. It is best in dappled-shade in rich, moist soil. There is an admirable white form, *T. hirta alba*, which occasionally has the odd spot of colour. In its native Japan it has been subject to much breeding and there are several cultivars. *T. hirta* 'Miyazaki' has cheerful gold variegated foliage.

Trillium

The wake robins are rhizomatous plants of which there are about 30 species in the family Trilliaceae/Liliaceae. They are native to North America and Asia. They derive their scientific name from the three-part division of foliage and flowers which they all have. Many gardeners have had difficulty in getting these lovely plants established. The best advice is to buy them from a specialist nursery run by people who understand their needs; never touch those miserable dried-out rhizomes, from dubious sources, sometimes seen in garden centres. They may be propagated by

division in the autumn, adding enriched soil. For more
intrepid gardeners, they may also be propagated by
seed, but it is unlikely that plants will flower in less
than four or five years. Most prefer acid soil but a few
demand alkaline.

Trillium chloropetalum
Origin: W. U.S.A.
Z: 6

❧ Some trilliums give off a pungent smell – one of
their common names is stinking Benjamin – but this
one has the most delicious, and unexpected, scent of
roses. The flowers, which appear in April, vary in
colour from deep red to white. The most beautiful of
all is a lovely pale rosy purple. The leaves, up to
6in/15cm across, are particularly decorative, with pale
veins and scatterings of dark marbling. They are rather
upright and slightly cupped, forming a graceful frame
for the exquisite flowers. A site that is at least semi-
shaded is needed, with moist, rich soil that is neutral or
alkaline. It is beautiful underplanted with smaller
bulbous plants that also have decorative foliage such as
erythroniums and *Cyclamen hederifolium*.

Trillium cuneatum
Origin: S.E. U.S.A.
Z: 6

❧ This is among the largest and stateliest of the trilliums with the fleshy stems rising as tall as 24in/60cm. The leaves are in proportion, with each of the three parts 4in/10cm long, making a bold shape, deeply veined with dark mottled marbling. The wedge-shaped flowers in April are a marvellous deep purple-black with an intricate pattern of almost black veins. This is one of the grandest of woodland plants and it is well worth lavishing on it the care which it demands. It needs humus-rich soil that never dries out but nor should it ever be water-logged. In the wild it is found in shady places in alkaline or neutral soil. A deep mulch of the best compost you can lay your hands on will help it to thrive. In the garden find a position that will provide shade when the trilliums are in leaf. A good place is underneath substantial spring-flowering deciduous shrubs such as corylopsis, hydrangeas, magnolias and viburnums.

Trillium erectum
Origin: E. North America
Z: 4

❧ The flowers of this trillium, which open in April, are among the most beautiful of the genus. They vary in colour but the very best are blood red etched with veins in an even deeper colour. The petals are rounded and pointed and their tips curve back sharply and the stamens are pale yellow. The foliage is rounded, a fresh mid green with pale veins, with a lustrous surface. It rises to a height of about 18in/45cm and has an air of the greatest distinction. *T. erectum albiflorum* has white flowers flushed with green and *T. erectum luteum* has a yellowish leaf stalk and rich red flower.

Trillium grandiflorum
Origin: E. North America
Z: 5

❧ This forms a burgeoning mound of foliage with profusely borne flowers. The leaves are especially attractive, deeply veined, twisting and creating a lively background to the flowers. These, appearing in April or May, are of a dazzling white, strikingly furrowed, set off by cheerful lemon-yellow stamens. It will form a shapely clump up to 18in/45cm high and must have a position that is at least partly shaded. It makes a wonderful underplanting to flowering shrubs; I have seen it beautifully used spreading underneath a

Illustration opposite:
Trillium cuneatum

Illustration:
Trillium grandiflorum

Magnolia stellata, producing its flowers as the last of the magnolia fade. The flowers are variable: there is an exceptionally beautiful double-flowered form, *T. grandiflorum flore-pleno*; and a rarely seen pink cultivar, *T. grandiflorum* 'Roseum', which is richly coloured with striking deeper-coloured veins. In its native habitat *T. grandiflorum* is found in limestone and is thus one of the few trilliums that will flourish in very alkaline soil.

Trillium ludovicianum
Origin: S.E. U.S.A.
Z: 6

❧ This is a rarity – but it is well worth seeking out. The leaves are particularly beautiful, up to 6in/15cm long, oval, with an undulating margin. They are especially handsomely marked with dark green marbling and a network of paler veins. The leaves are held well up on stems up to 15in/35cm tall. The flowers are a lovely glowing, rich red with green sepals striped with red – they have a vague whiff of dirty socks. It flowers in April and makes one of the most ornamental of all its tribe. The oval fruit are a striking pale purple

colour. In the wild it grows in rich, moist soil in forests or shady places by the banks of rivers. If you can find a similar position in your garden there are few plants more worthwhile to grow.

Trillium ovatum
Origin: W. North America
Z: 5

❧ This beautiful trillium has especially elegant flowers well carried above spreading foliage. They appear in April or May, with petals separate, white, with undulating margins and a flush of rosy pink. As the flowers age the pink begins to suffuse the whole flower, with deeper coloured veins becoming more visible. They are held on 4in/10cm red-brown fleshy stems above the foliage which is low lying, spreading across the surface of the ground. In the wild this is a plant of coniferous forest and a shady position will suit it best in the garden. In appropriate conditions it makes admirable floriferous ground cover, the foliage remaining attractive long after the flowers have gone.

Trillium pusillum
Origin: E. U.S.A.
Arkansas, Missouri, Texas
Z: 6

❧ This little trillium is quite unlike any other species. It stands very upright with flowers and sepals at the tip of fleshy stems which are up to 12in/30cm high. The base of the stems is a dark purple, becoming green farther up. The flowers in April are white, or the palest pink, with prominent pale lemon yellow anthers. In the wild it is found on dry wooded slopes and in low woodlands of the coastal plains. Plants from the western states are slightly larger than those from the east which are sometimes seen listed as *T. pusillum* var. *virginicum*. I have seen it grown very well on scree in a rock garden. It would be a lovely ornament of a shady, woodland area where it should be planted in a bold clump – the flowers, which are rather small in relation to the height of the plant, look best in a group.

Trillium rivale
Origin: W. U.S.A.
Z: 4

❧ This is one of the smaller trilliums, with flowering stems no more than 4in/10cm long. The flowers are variable in colour from white to pale pink – the most desirable are handsomely freckled within with purple spots. Like all trilliums the flowers have the three-part division of petals but here they are so neatly overlapped as to form a cupped circle. The leaves, on the other hand, are widely separated, rounded and pointed, with the flowers held well above. Despite its exquisite delicacy *T. rivale* is found in harsh conditions in its native habit – in mountainous regions as high as 4,000ft/1,200m and in coniferous woodland. In the garden it will do best in part-shade in rich soil. It grows particularly well in the cooler gardens of Scotland. It is one of those plants whose greatest virtue is its own distinctive beauty rather than in its ability to mix with others. It is an admirable rock-garden plant and I have also seen it looking very beautiful spreading out at the foot of a strawberry tree, *Arbutus unedo*.

Trillium sessile
Origin: N.E. United States
Z: 4

❧ Sessile means stalk-less which in this case refers to the flower which sits neatly with its base hard against the foliage underneath – like something delicious sitting on a plate. The flowers appearing in April are a dazzling apparition – a rich deep maroon with a glistening surface and framed in green-purple outer petals. They are shown off to marvellous advantage by the leaves which are up to 4in/10cm across, pale green but richly mottled with darker green markings, creating a sumptuous effect. The whole plant stands no more than 12in/30cm high. In the wild it is found in alkaline soil and like other trilliums needs moist humus-rich soil in the shade. It is wonderful with Lenten hellebores (*Helleborus orientalis*) whose flowers will be fading just as the trilliums come into full beauty. The cultivar *T. sessile* 'Rubrum' has resplendent flowers of a more crimson colour.

Triteleia

There are about 15 species of triteleia, all corms, in the family Alliaceae/Liliaceae. The genus has been a botanist's battlefield with species transferred from one genus to another. Some species of *Triteleia* are now *Ipheion* or *Dichelostemma*.

Illustration: *Triteleia* 'Koningin Fabiola'

Triteleia laxa
Origin: California, Oregon
Z: 7

❧ Previously known as *Brodiaea laxa*, this summer-flowering plant is of the kind that illuminates odd corners of the garden, adding just those decorative details that give character. It flowers in June or July, with umbels of upward-pointing little trumpets of violet-blue or white. Each flower is carried on a long stalk and the whole umbel is held on a wiry stem about 12in/30cm high. It takes up practically no ground space and the flowers rise above lower plantings and dance in the breeze. It thrives in quite poor soil and needs a sunny position. In these conditions it will seed itself happily (but not *too* happily). It may also be propagated by dividing clumps of corms in early spring. Corms should be planted 4in/10cm deep. It is an excellent front-of-the-border plant which may also be planted in places too small to allow other plants. An admirable cultivar, *T. laxa* 'Koningin Fabiola' (or 'Queen Fabiola'), has rich coloured purple-blue flowers borne in great profusion.

Tropaeolum

There are about 80 species of tropaeolum in the family Tropaeolaceae, all native to South America. Among them is the common annual nasturtium, *T. majus.*

Tropaeolum speciosum
Origin: Chile
Z: 7

&. This enchanting plant is well-known to be difficult to establish and, once established, to wander in unpredictable places. It seems to prefer cool, damp places and in Britain is always seen at its best in the north of England and in Scotland. It is a rhizomatous climbing plant which scrambles through other plants casting a veil of dazzling scarlet flowers in June or July wherever it goes – rising to at least 10ft/3m. The flowers are like diminutive nasturtiums, trumpet-shaped with a sharp spur behind. They have yellow veins at the base and yellow stamens, and are carried on plum-coloured stems. The flowers are followed by rich blue berries. The foliage is also decorative, with glaucous lobed leaves. It is most commonly seen on yew hedges where the dark green makes a superb background. Although most frequently seen in gardens of acid soil it will grow perfectly well, with plenty of humus, in calcareous soil. It seems to establish itself best in a shady, rather dry place such as at the base of a yew hedge. It has a natural tendency to establish itself on the cooler side of its host plant. It may be propagated by dividing rhizomes in the spring.

Tropaeolum tuberosum
Origin: Bolivia, Peru
Z: 8

❧ This tuberous climbing nasturtium has a splendid combination of orange-yellow flowers and glaucous grey leaves. The flowers are similar to those of *T. speciosum* but neater. They are warm orange on the outside and yellow-orange within, opening in late summer and continuing well into the autumn. The attractive lobed leaves appear well before the flowers. It must have a sunny position and is in any case not a plant for cold gardens. It will flower best scrambling through other plants on a south-facing wall where it will rise to at least 10ft/3m. It should have rich, moist soil. It is a marvellous climber to provide flowering ornament in a season when most clematis and roses have ceased to flower. It looks beautiful among the glistening seed-heads of *Clematis orientalis*.

Tulbaghia

Named after an 18th-century Governor of the Cape, Ryk Tulbagh, these attractive bulbs and rhizomes are found only in South Africa where they are known as wild garlic. There are about 25 species, in the family Alliaceae/Liliaceae. Most of the species need a frost-free climate but some, from high mountain sites, may be hardy in temperate gardens in Zone 8. Failing that, they make marvellous plants for pots and troughs, with the added bonus that some species are deliciously scented.

Tulbaghia violacea
Origin: South Africa
Z: 8

❧ This rhizome produces slender glaucous grass-like stems crowned with dazzling umbels of flowers. The flowers in July or August are very elegant – little elongated trumpets, a warm lilac-violet, opening out abruptly, with deeper stripes down the middle of the petals. Each umbel carries up to fifteen flowers which

are sweetly scented. The flowering stems, 24in/60cm high, rise above narrow glaucous leaves – which smell strongly of onion when bruised. This is a plant to be admired for its exquisite detail – form, scent and colour – rather than for its structural presence in a busy border. It must have very sharp drainage and the sunniest position you can find. It is marvellous by itself, in an appropriate little bed or in a pot or trough, or with a carefully chosen partner. I have seen it looking magnificent with the vermilion-flowered *Zauschneria californica* which flowers at the same time. *T. violacea* 'Silver Lace' is a handsome cultivar with variegated leaves, which are grey with a fine edging of silver on each side. It is a most elegant plant, the silver setting off the violet to especially good effect.

Tulipa

The name tulip comes from the Turkish for a turban, *tulbend*, which the flower somewhat resembles. There are about 100 species, in the family Liliaceae/Liliaceae, from the Mediterranean region and Central Asia. Among the first garden plants to be given cultivar names, they have excited the interest of gardeners at least since the 16th century and in the Low Countries in the early 17th century they became the subject of frenzied investment speculation. The market

crashed but tulips have remained a Dutch speciality and essential garden plants.

These are the most precious spring-flowering bulbs, with a wide range of colours and shapes. Their flowering period runs through three months, from April to June when the beautiful *T. sprengeri* ends the season with a splendid trumpet blast of brilliant colour. In the garden they have many uses – in the border or meadow garden, in the rockery, alpine trough or other containers. Many make easy association plants and the colours available are so diverse that very specific effects are possible. The more exotic kinds, those with elaborate petals richly splashed or striped with contrasting colours, demand a simple setting.

The vast majority of tulips seen in public places and in gardens are cultivars, so remote from their wild origins that nothing is known of the species from which they derive. I describe below in some detail some of the species many of which – apart from the beauty of their flowers – have the great advantage that they will naturalise in conditions that suit them; almost all modern cultivars are sterile. The cultivars will usually not flower well from year to year unless they are lifted after flowering, stored in a dry place and replanted in the following autumn, even then they will require a further year to produce full-sized flowers.

Illustration:
Tulipa 'Candela'

Those left to struggle, particularly in heavy soil, may
rot, are prey to slugs, and the few survivals often
produce irritatingly sickly flowers. Many gardeners
give up the struggle and regard them as annuals. Others
claim that, with good drainage and plenty of lime, they
will flourish for years. In my own heavy, wet soil I
have never achieved this except in a large old copper.
Here, with excellent drainage, a group of the yellow
lily-flowered 'Candela' has formed a successful colony,
flowering regularly for ten years.

Bulbs should be planted between 6in/15cm and
8in/20cm deep, the smaller bulbs (i.e. species) more
shallowly than the larger ones. When planting them
avoid at all cost any hint of regimentation unless you
are planning a severely municipal effect. It will help to
achieve naturalistic drifts if you arrange the bulbs on
the ground before planting. Late autumn or early
winter is the recommended time of planting but some
gardeners have reported successes with tulips planted as
late as February. If you are planning to encourage a
permanent colony, leaves should be left after flowering
as they are the channel through which the new bulb
receives nourishment. The dying leaves are ugly but
may be concealed by the emerging new foliage of
neighbouring herbaceous plants.

Tulips have been divided by the Dutch Koninklijke

Illustration: *Tulipa* 'White Triumphator'

Algemeene Vereening voor Bloembollenculture into four groups – Early Flowering, Mid-Season, Late Flowering and Species and their Hybrids. These broad divisions have been broken down into fourteen sections according to the general shape or character of the flowers. The group Species and their Hybrids includes tulips that flower early, mid and late. Those derived from *T. kaufmanniana*, with their characteristically purple-mottled leaves, such as 'Glück', with yellow and scarlet striped flowers, are among the earliest to flower. Hybrids and cultivars of *T. fosteriana* give one of the best yellow single tulips, 'Candela', which flowers in early April; 'Purissima', an excellent white double of the same origin, flowers in early May. The third group of species, derived from *T. greigii*, all flower in the mid season. They have in common purple-marked leaves and rather short, stubby flowers. Most have orange-red flowers and some are striped with yellow. 'Red Riding Hood' is an attractive scarlet one with goblet shaped flowers. The

Illustration:
Tulipa 'Apricot Parrot'

late flowering Lily-Flowered Group, which flowers at the beginning of May, provides some of the most beautiful cultivars. They have in common long, elegantly-shaped flowers with pointed petals. 'China Pink' is a good warm pink; 'White Triumphator' is one of the best of all the single whites, eventually opening its petals into widely-spreading wings; 'West Point' is a fine clear yellow with strikingly pointed petals which reflex backwards. The Viridiflora Group, as the name implies, have flowers flushed with green: 'Spring Green' flowering in early May has pale yellow flowers suffused with lime-green making it a particularly good ingredient of a white, yellow and green arrangement. 'Mount Tacoma', flowering at the same time, is a fine double-flowered white. The Parrot Tulips, all late-flowering, evoke the most lavishly exotic blooms of 17th-century tulipomania. Some have flower-heads so large and heavy as to cause them to flop over. Many are irresistibly over the top – such as 'Apricot Parrot' with pleated and frilled petals suffused with apricot, pink and cream like some stupendous sundae. 'Black Parrot', on the other hand, is an exceptional Satanic purple of the deepest colouring, wonderful with pale grey foliage – I have seen it lookinmg superb among the silver foliage of cardoons, *Cynara cardunculus.*

Tulipa clusiana var.
chrysantha
Origin: Garden
Z: 5

ঌ *Tulipa clusiana* is named after the great botanist Carolus Clusius who, in the second half of the 16th century, was one of the first to study bulbs systematically. The flowers of the variety *T. clusiana* var. *chrysantha* open in April and are a rich golden yellow (rather than the white of *T. clusiana*) and marked on the back of the petals with red smudges. It rises to about 10in/25cm tall, and the newly opened flower has upright heads with crisply pointed petals but as it ages the flower heads flop and the petals twist creating a splendidly languid effect. It is best in sun in sharply drained soil where many gardeners have found it will establish itself well.

Tulipa kaufmanniana
Origin: Central Asia
Z: 5

ঌ This March-flowering tulip is a marvellous spring plant. The flowers are creamy yellow with a rich golden centre and in full sunshine they open very widely, with the petals quite separate. The backs of the petals are marked with red and the foliage is especially attractive with broad, glaucous green, undulating leaves. As many as five flowers are carried on each stem which rises to a height of 8in/20cm. In the wild it is found in dry, stony places, quite high up. In the garden it should have a sunny, well-drained site where, in its summer dormancy, it can dry out completely. The front of a south-facing border, covered later in the season by the foliage of other plants, is a good place.

Tulipa linifolia
Origin: Central Asia
Z: 5

꽃 This little tulip, rising no more than 6in/15cm high, has an exceptionally distinguished character. The leaves are narrow, folded down the centre, a lustrous green edge with fine red margins. The flowers start to open when they are still shrouded by the foliage, showing a dazzling scarlet. When they open fully in May the pointed petals spread sideways giving a more dishevelled appearance and revealing black splotches at the base of the petals. In the wild it is found high up in the mountains in dry, stony areas. In the garden it makes an excellent front-of-the-border ornament in a sunny place with excellent drainage. Here it may well settle down and form a self-perpetuating colony. The colour associates especially well with silver-leafed shrubs of Mediterranean character such as lavender, santolina or small-leafed sage.

Tulipa marjolettii
Origin: S. Europe
Z: 6

꽃 This is a valuable little tulip which will settle down and multiply with ease. It produces its flowers in early May, a warm creamy yellow with variable red edgings to some of the petals. As the flowers age they take on the colouring of a ripe yellow peach. The flower remains shapely, never spreading its petals fully open. It will grow 9in/23cm high and makes a good companion for other low-growing plants that perform in the same season such as the smaller bearded irises (especially pale blue ones), forget-me-nots and the chalky-blue *Veronica gentianoides*. I have seen it looking marvellous growing through the finely cut foliage of *Dicentra formosa* 'Stuart Boothman' whose pale salmon coloured flowers make a pretty accompaniment. Although some authorities say that it needs a sunny position and sharp drainage, I grow it successfully in half-shade in heavy soil.

Tulipa orphanidea
Origin: Balkan Peninsula
Z: 5

❧ The colour of this tulip is as close as nature gets to producing an orange flower. The flowers, which appear in late April or May are at first rounded and pointed but open out into a cupped shape, resembling a miniature water-lily. The colour is variable from red-brown to a marvellous rusty-red but always with a soft stripe of yellow on the inside of the petals and a yellow flush to the outside. The centre of the flower is black with deep purple anthers. This is one of the prettiest of all the wild tulips, with its flowers held high on thin stems 6in/10cm long. The leaves are a pale, glaucous grey. In the wild it is found in damp meadows and rocky places in the mountains. In the garden it should have a sunny position in humus-rich soil. *T. whittallii* is either a synonym or a slightly shorter version of exactly the same thing.

Tulipa praestans
Origin: Central Asia
Z: 5

❧ In my garden this is the first tulip to flower, opening in late March or April, and providing an exhilarating jolt of red when everything else in the garden seems green, yellow or blue. The single flowers are held on stems 7in/18cm tall, opening to show a splendid rich scarlet, with decorative glaucous grey foliage below. A slightly larger cultivar, *T. praestans* 'Van Tubergen's Variety', has a hint of orange in the scarlet. It should have a well-drained position in the sun and it looks especially beautiful against the silver-grey foliage of such small shrubs as santolina. The foliage is a decorative glaucous grey.

Tulipa saxatilis
Origin: Crete, Turkey
Z: 6

❧ The flowers of *T. saxatilis* open in April, pale lilac pink with a striking yellow smudge at the base of each petal – as though dipped in gold-dust. The petals are pointed and the flower stems are 8in/20cm long. It is one of the most striking and elegant of the species

tulips. *T. bakeri* is similar but smaller, only 6in/15cm high, and with a deeper lilac-purple colour of flower which somewhat resembles a tall crocus. Both these demand a warm, sunny position and excellent drainage.

Tulipa sprengeri
Origin: Turkey
Z: 5

ૐ Flowering well into June *Tulipa sprengeri* gives the tulip season a triumphant send off. It is among the most decorative of all, a dazzling blood-red flower carried on a stem up to 12in/30cm high. The petals are long and pointed with a hint of pale yellow on the outside and brilliant yellow anthers within. The foliage is a fresh green with a lustrous texture. This is one of the tulips which gardeners find most easy to naturalise. It will seed itself with abandon – indeed some lucky gardeners have complained of having too much of it. It is said to do well in sun or shade, in peat or dry soil. I have seen it in a sunny courtyard having sown itself liberally along the cracks between paving stones. It looks wonderful in a sunny border with small shrubs like cistus, lavender and thyme. I have seen it brilliantly planted with the silver-pink cistus 'Peggy Sammons' among whose lower branches it threaded its way.

Tulipa sylvestris
Origin: European
Mediterranean
Z: 5

❧ This European wild tulip is of slightly mysterious origin. It is now fairly wildly naturalised in many lowland areas. It is a marvellous garden plant for it has much of the brilliance of the cultivars with all the charm and grace of the species. It flowers in April, the flower buds swaying on tall, thin stems up to 12in/30cm high, marked with a slight bronze tinge. The flowers when fully open are a sprightly yellow, sweetly scented and the petals are rounded and pointed – some of them curling backwards gracefully. It is a good tulip for naturalising, in sun or semi-shade. Plant it under spring-flowering deciduous shrubs and trees such as *Viburnum carlesii* or *Amelanchier canadensis*.
T. sylvestris ssp. *australis* is almost identical but on a much more delicate scale, growing no more than 9in/23cm tall and with narrower flowers whose petals are smudged with green on their backs. It has very slender stems which allow the flowers to sway elegantly in the slightest breeze.

Tulipa tarda
Origin: Central Asia
Z: 5

&⬥ Despite its name this is one of the earlier flowering tulips. The flowers open in April, white with a lemon centre. As many as eight flowers are produced on each bulb and a good clump gives a wonderful impression of floriferous abundance with flowers packed together among the distinguished glaucous green foliage. The buds are very elegant before opening, long and pointed with a smudge of green down the back of each petal. The flowering stems are no more than 6in/15cm long. This is a good tulip for naturalising in a sunny, rather dry place. I grew it successfully on a south-facing slope, where in a warm spring it would flower as early as March. Here it intermingled with *Crocus tommasinianus* and stars of Bethlehem (*Ornithogalum umbellatum*) making a lovely spring mixture.

Uvularia

Uvularia grandiflora
Origin: S.E. U.S.A.
Z: 4

There are about five species of uvularia, all rhizomes, in the family Convallariaceae/Liliaceae.

&⬥ Any gardener who can provide the right conditions for this lovely plant should do so. Its stems thrust through the earth in the early spring looking very much like those of its close relation, Solomon's Seal. The leaves are pale green and lightly furrowed. The flowers which open in March or April are a beautiful

crisp lemon yellow, hanging downwards, up to 2in/5cm long, and half-concealed among the foliage. The petals twist slightly and their bases are striped with green. A colony of *Uvularia grandiflora* makes a dazzling sight, with the stems rising as high as 30in/75cm and the mass of twisting leaves creating a lively pattern. In the wild it grows in light acid soil but it will grow in neutral soil. It is at its best in shade or part-shade and makes a wonderful underplanting for larger shrubs in the naturalistic setting of a woodland garden. It is also an admirable companion for other herbaceous plants, such as ferns, hellebores and pulmonarias which need the same conditions. It may be propagated by dividing the rhizomes in autumn.

Veratrum

Veratrum album
Origin: Europe, N. Africa,
Siberia
Z: 5

Illustration opposite:
Veratrum album

The name comes from the Latin for 'truly black' – a reference to the plants' dark rhizomes. There are about 25 species, in the family Melanthaceae/Liliaceae, all native to areas with temperate climates.

❧ The most dazzling quality of the false helleborine is its resplendent foliage which unfurls from the ground in April, curved and intricately pleated like rare and wonderful fabric. These are up to 12in/30cm long, gracefully rounded and coming to an unexpected point. The flowers merge in June, tall racemes of green white flowers, forming a stately spire of green and cream up to 24in/60cm high which would be more impressive if the leaves were not so striking. These form a beautiful background for other herbaceous plantings in the spring – tulips and primulas are good companions. It should be given a sunny site in rich moisture-retentive soil. Later in the summer the leaves lose their freshness and become rather tatty. If possible try and plant it so that other plants – perhaps geraniums or the larger campanulas – will grow up to conceal this defect. It is best propagated by dividing the rhizomes in the autumn; seeds germinate easily but it takes four or five years to produce flowering plants.

Zantedeschia

Zantedeschia aethiopica
Origin: South Africa
Z: 8

There are five species of arum lily in the family Araceae. They are all rhizomes, native to southern Africa.

❧ This is one of the less tender arum lilies which makes a striking ornamental plant. The foliage is magnificent, large loosely arrow-shaped leaves of a dark green colour, with a glossy surface and undulating margins. The flowers, with white spathes shaped like irregular trumpets, open in June from curious twisting lime-green buds. It will grow to about 4ft/1.2m high. In growth it needs much water and is at its best in a shady position in very rich soil. In less favoured climates it makes a superlative pot plant but it will need very rich, moisture-retentive compost and watering at

least daily in warm weather. It may be propagated by removing suckers in spring. *Z. aethiopica* 'Crowborough' is almost identical but will flourish in much drier conditions. *Z. aethiopica* 'Green Goddess' has striking mysteriously green spathes and is hardier than the type.

Zephyranthes

Zephyranthes candida
Origin: South America
Z: 8

There are about 70 species of zephyranthes in the family Amaryllidaceae/Liliaceae.

❧ Flowering towards the end of August, usually when the weather becomes cooler, this very attractive bulb is a welcome surprise. The flowers are white, star shaped with upright pale yellow anthers. The petals have pale grey stripes down their length and their tips are occasionally splashed with pink – which may also be visible on their underside. They are held at the tips of stiff stems, rosy pink where the flower starts. The leaves are very thin, grass like, rising to 12in/30cm, taller than the flowers. In the wild it is found in marshland and in the garden it is said to do best in rich, moist soil. I have seen it flourishing in a rather dry place at the front of a border – at all events, it needs sun.

Bulbs for Different Sites

Bulbs for Shade
Many species of bulbous plants have woodland or shady places as their natural habitat. In the garden they are especially valuable, ornamenting places which may otherwise be difficult to plant.

Allium triquetrum
 A. ursinum
Anemone blanda
 A. nemorosa
Arum italicum ssp *italicum*
 'Marmoratum'
Cardiocrinum giganteum
Convallaria majalis
Corydalis flexuosa
Cyclamen pupurascens
Disporum flavens
Erythronium species
Fritillaria pallidiflora
Galanthus species and cultivars
Hyacinthoides non-scripta
Iris douglasiana and cultivars
Leucojum vernum

Lilium martagon
 L. monadelphum
Nomocharis pardanthina
 punctulata
Polygonatum × *hybridum*
Ranunculus ficaria
Sanguinaria canadensis
Smilacina racemosa
Trillium species
Uvularia grandiflora

Bulbs for Different Sites

Illustration:
Zephyranthes candida

Bulbs for Positions in Full Sun
Many bulbs demand a sunny, protected position. In some cases they will do well in quite poor, thin soil but others need both sunshine and a rich, moist soil.

Albuca nelsonii
Alstroemeria haemantha
 A. psittacina
× *Amarcrinum memoria-corsii*
Amaryllis bella-donna
Asphodeline lutea
Asphodelus aestivus
Cosmos atrosanguineus
Crinum bulbispermum
 C. × powellii
Crocus gargaricus
Dahlia species and cultivars
Dichelostemma congestum
 D. ida-maia
Dierama dracomontanum
 D. pendulum
 D. pulcherrimum
Eremurus robustus
Fritillaria assyriaca
 F. michailovskyi
Galtonia species
Geranium tuberosum
Gladiolus callianthus
 G. papilio
 G. 'The Bride'
Iris missouriensis
 I. orientalis
 I. unguicularis
Ixiolirion tataricum
Kniphofia species and cultivars
Libertia formosa

Lilium candidum
 L. longiflorum
 L. speciosum
Moraea huttonii
Nerine bowdenii
Ornithogalum arabicum
Paradisea lusitanica
Rhodohypoxis baurii
Romulea nivalis
Schizostylis coccinea
Tulbaghia violacea
Zephyranthes candida

Bulbs for damp sites
Although many bulbs need rich moisture-retentive soil some are at their best by the banks of streams or pools with the water lapping at their stems

Canna species and cultivars
Dactylorhiza elata
 D. praetermissa
Epipactis palustris
Iris ensata
 I. laevigata

I. orientalis
I. pseudacorus
I. sanguinea
I. sibirica
Zantedeschia aethiopica

Bulbs for Meadows and Orchards

Many bulbs are at their most decorative in naturalistic settings. In meadows or orchards with grass, the important thing is to leave plants until they have set seeds or the foliage has withered before cutting the grass. This may be too late for many people's taste – but the effect is marvellous.

Anemone blanda
Camassia quamash
Cyclamen coum
Fritillaria meleagris
   F. pyrenaica

Hyacinthoides non-scripta
Iris latifolia
Narcissus species and cultivars
Ornithogalum pyrenaicum
Scilla liliohyacinthus

Bulbs for Pots and Containers

Bulbs are in many ways the perfect pot plants. They are easy to move when dormant and, for the trickier kinds, the pot makes it easy to provide exactly the right position and soil.

Albuca nelsonii
× *Amarcrinum memoria-corsii*
Amaryllis bella-donna
Cosmos atrosanguineus
Crinum × *powellii*
Eucomis species
Fritillaria persica
Galtonia candicans

   G. regalis
Gladiolus callianthus
   G. papilio
   G. 'The Bride'
Iris species
Lilium species and cultivars
Narcissus species and cultivars
Tulipa species and cultivars

Bulbs with Specific Qualities

Bulbs of Architectural Character

These are bulbs which by virtue of bold shapely foliage or spectacular flowers will make a structural contribution to the garden.

Albuca nelsonii
Allium aflatunense
   A. christophii
   A. nigrum
   A. rosenbachianum
   A. schubertii
× *Amarcrinum memoria-corsii*
Amaryllis bella-donna
Anthericum liliago

Asphodeline lutea
Asphodelus aestivus
   A. albus
Camassia leichtlinii
Canna species and cultivars
Cardiocrinum giganteum
Crinum × *powellii*
Crocosmia species and cultivars
Dactylorhiza elata

Plants of Architectural Character (continued)

Dierama pulcherrimum
Dracunculus vulgaris
Eremurus robustus
Eucomis species
Fritillaria imperialis
  F. persica
Galtonia candicans
  G. regalis
Gladiolus communis ssp.
  byzantinus
Hedychium coronarium
Iris magnifica
  I. orientalis
  I. pseudacorus
  I. sanguinea
  I. sibirica
Kniphofia caulescens
  K. uvularia 'Nobilis'
Lilium candidum
  L. longiflorum
  L. martagon
  L. monadelphum
  L. pardalinum
Moraea huttonii
Nectaroscordum siculum

Paradisea lusitanica
Polygonatum × hybridum
Sisyrinchium striatum
Veratrum album

Bulbs with Specific Qualities

Bulbs for Scent

Some bulbs have among the most delicious scent of all plants. Several of those listed here are also suitable for cultivation in pots or containers, making them especially good for terraces or other sitting places were the scent will be particularly appreciated.

Albuca nelsonii
× *Amarcrinum memoria-corsii*
Amaryllis bella-donna
Cardiocrinum giganteum
Convallaria majalis
Cosmos atrosanguineus
Crinum bulbispermum
  C. × *powelli*
Crocus tommasinianus
Cyclamen repandum
Fritillaria michailovskyi
Gladiolus callianthus
Hedychium coronarium
Hemerocallis citrina
  H. lilioasphodelus

Hermodactylus tuberosus
Iris orientalis
  I. unguicularis
Lilium auratum
  L. candidum
  L. longiflorum
  L. regale
  L. speciosum
Muscari macrocarpum
Narcissus jonquilla
  N. papyraceus
  N. poeticus
  N. tazetta
Smilacina racemosa
Tulbaghia violacea

Bulbs with Decorative Foliage

Several bulbs have ornamental foliage, continuing to be ornamental long after flowers have faded. In some cases the foliage becomes more attractive after flowering; in other cases, such as the beautiful *Arum italicum* ssp. *italicum* 'Marmoratum', the foliage is the most beautiful feature.

Allium karataviense
Arum italicum ssp. *italicum*
  'Marmoratum'
Canna species and cultivars
Convallaria majalis
Corydalis flexuosa
  C. lutea
  C. solida
Crinum species
Crocosmia species and cultivars
Cyclamen hederifolium
  C. purpurascens
  C. repandum
Dahlia species and cultivars

Dracunculus vulgaris
Eranthis hyemalis
Erythronium species
Iris pseudacorus
Polygonatum × *hybridum*
Ranunculus ficaria
  'Brazen Hussy'
Sanguinaria canadensis
Smilacina racemosa
Trillium species
Tropaeolum species
Uvularia grandiflora
Veratrum album
Zantedeschia aethiopica

Hardiness Zones

Temperature Ranges		
F	Zone	C
below −50	1	below −45
−50 to −40	2	−45 to −40
−40 to −30	3	−40 to −34
−30 to −20	4	−34 to −29
−20 to −10	5	−29 to −23
−10 to 0	6	−23 to −17
0 to 10	7	−17 to −12
10 to 20	8	−12 to −7
20 to 30	9	−7 to −1
30 to 40	10	−1 to 5

Hardiness zones are based on the average annual minimum temperature in different areas, graded from Zone 1, the coldest, to Zone 10, the warmest; thus, if a plant has the rating Zone 7 it will not dependably survive in a zone of a lower number. But the data are only broadly relevant and are more valid for continental climates than for maritime ones. In Britain and many parts of Europe, for example, local microclimate rather than the hardiness zone band is more likely to determine a plant's hardiness. It should also be said that a plant's chances of survival may be influenced by other things than temperature; drainage, rain, amount of sunshine and protection from winds may make a fundamental difference.

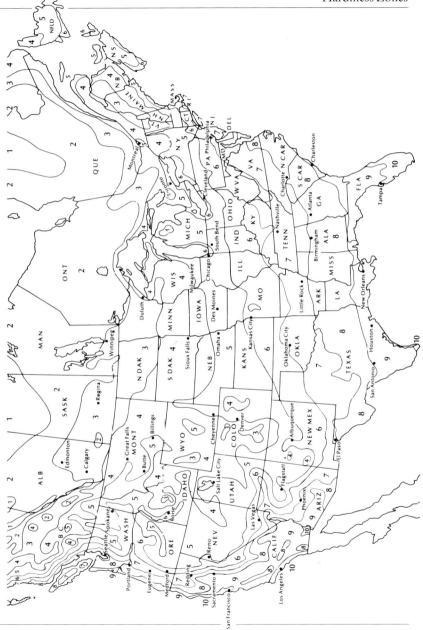